Middle
School
Math

Interventions

Dealing with the Difficult Concepts

Dr. Carl Seltzer

Printed in the United States of America.

This book is printed on recycled paper.

Order Number 211224

ISBN 978-1-58324-332-9

A B C D E 15 14 13 12 11

395 Main Street
Rowley, MA 01969
www.didax.com

Dedication and Appreciation

This book is dedicated to my wife, Charlotte, who helped me put it together and graciously allowed me vacation times to work on it.

Special thanks go to Maggie Holler, who did the initial editorial work and offered so many valuable suggestions, and to Dr. Bruce Phillips for giving me several of the ideas that I used in Chapter 5.

Also, many thanks to Linda Levine, Mathematics Supervisor in Osceola County, Florida, for giving me the idea to write this book and for helping me solicit teachers in her county to participate in the original study. I am indebted to the Osceola County teachers for their valuable suggestions contributing to the content.

Also, to those hundreds of other teachers who also submitted their ideas for the topics included in these pages, thank you.

Dr. Carl Seltzer

Table of Contents

Introduction

Teachers generally identify certain topics in mathematics as being more difficult to teach than others. Many students struggle with these topics, becoming frustrated and sometimes, unfortunately, developing a lifelong aversion to mathematics. Traditional ways of addressing these concepts may not be effective for some students.

To identify the most difficult topics in mathematics, I surveyed several hundred teachers of students in grades 5 through 9. The six topics that teachers identified most frequently (in order) were:

1. Fractions

2. Word problems (story problems)

3. Division (long division)

4. Multiplication (especially facts)

5. Algebraic thinking

6. Decimals

This book addresses each of these topics by reteaching foundational concepts using a step-by-step approach. Each topic occupies a chapter in this book in a developmental sequence rather than the order of frequency of response.

Important Ideas to Consider

Children may not be successful in learning mathematics for many reasons. Current research has identified how children learn mathematics, which allows us to see where they might derail. These learning trajectories are included at the beginning of each chapter in this book, along with background information on each topic and what research says about student difficulties. Although some of the problem can be attributed to students' lack of experience in mathematics, motivation, or perseverance, much of the problem lies in the way mathematics is taught. Following are some ideas for math instructors to consider.

Mathematics Is a Science

What is mathematics? I believe one's definition of mathematics is important because we tend to teach math depending on what we think it is. I've read and heard many definitions, but I think the following is the best:

Mathematics is the science of patterns and relationships.

That's right—mathematics is a science! So if it's a science, shouldn't it be taught as a science? A scientist does many things, but all scientists do four basic things:

1. Perform explorations, investigations, and experiments (all basically the same thing).

2. Make observations.

3. Draw conclusions (theorize).

4. Perform verifications.

When you think about it, shouldn't these four things be present in our mathematics classes? I believe so—maybe not every day, but surely they should be the dominant feature of most classes.

The Role of Memory

If you teach something on Monday and your students have forgotten it by Friday, I believe you have wasted both your time and theirs. Memory plays an important role in learning mathematics, but what is the best way to present a mathematics lesson so that it will be remembered?

To answer this question, let's look at some research on memory. In 1969, Edgar Dale conducted a famous study that resulted in some findings with which teachers may be all too familiar. This study, known as the Mobil Oil Study, found that the method of presentation had a significant impact on how long information was retained. The following chart identifies the method of presentation and the percentage of the material that was remembered just 72 hours later.

Cone of Learning
We Tend to Remember Our Level of Involvement

(developed and revised by Bruce Hyland from material by Edgar Dale)

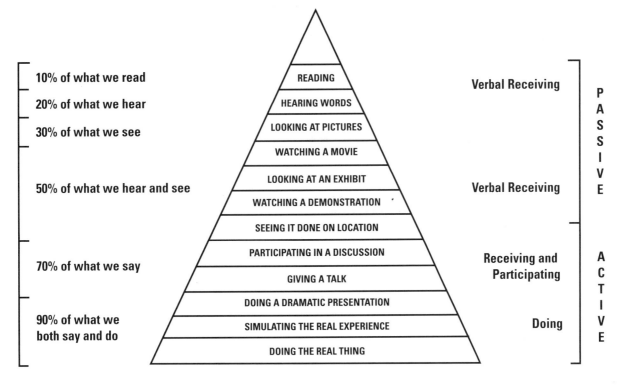

Edgar Dale, Audio-Visual Methods in Teaching (3rd Edition). Holt, Rinehart, and Winston (1969).

Dr. Lola May, the late distinguished mathematics educator, once said, "There are three things to remember when teaching: know your stuff, know whom you are stuffing, and then stuff them elegantly." In other words, we need to teach toward that 90 percent.

Thinking Skills

Another important aspect of teaching mathematics is to teach toward the higher-level thinking skills. Years ago, Dr. Benjamin Bloom conducted extensive research regarding thinking skills and formulated what became known as "Bloom's Taxonomy." Listed below are the Bloom's Taxonomy Thinking Levels translated into mathematical terms. They are listed from lowest-level thinking to highest.

1. Recognition

2. Facts

3. Algorithms

4. Structure

5. Modeling

6. Analysis

7. Decision making

While all of these skills are important, Bloom postulated that the higher the level of skill, the greater the achievement and the higher the productivity. Recent studies have tended to verify Bloom's results.

Phases of Teaching Mathematics

Another aspect of teaching mathematics is the recognition that mathematics learning involves four basic and distinct phases that need to be taught in the following sequence:

1. Concept development

2. Skill development

3. Problem solving

4. Applications

Problem solving and applications are similar but not exactly the same. Applications generally apply to real-world problems (for example, "What would be the price of a $10 item if it was marked 30% off?"). Problem solving is a prerequisite for applying mathematics to such problems. However, not all problem solving has real-world applications. For example, it might be necessary to find the 65th term in the sequence 3, 5, 7, 9, … to solve a real-world problem, but probably not.

Some educators argue whether concept and skill development or the ability to solve problems is most important in math learning. I would argue that all of the phases are critical to understanding mathematics and need to be incorporated into any mathematics lesson.

Motivation and Discovery

Effective learning takes place when students are (a) self-reliant, (b) self-confident, (c) willing to take risks, (d) organized, and (e) motivated. However, the most important of these is motivation.

Motivation comes from many sources, but I believe one of the best sources is the sense of discovery that students experience when they grasp a new concept. Alfred North Whitehead once said, "From the beginning of his education, the child should experience the joy of discovery." Discovery creates excitement, stimulates interest, and yields success. Success motivates.

Every once in a while we need to reinvent the wheel. Not because we need lots of wheels, but because we need lots of inventors.

—*Author unknown*

In 1970, Piaget wrote: "Each time one prematurely teaches a child something he could have discovered for himself, that child is kept from inventing it and consequently from understanding it completely." We need to try a discovery approach to teaching. It pays a huge dividend.

Finally

This book presents models for reteaching the math concepts that are the most difficult for middle school students to comprehend. The models show how one might approach the teaching and understanding of specific mathematical ideas through discovery and the recognition of patterns. I have purposely not presented lots of problems to be solved. Instead, I encourage teachers and students to provide their own examples, as needed, for mastery.

Dr. Carl Seltzer

Correlation to Common Core State Standards

	Ch. 1: Multiplication	Ch. 2: Division	Ch. 3: Fractions	Ch. 4: Decimals	Ch. 5: Algebra / Mental Math	Ch. 6: Word Problems
Grade 4						
Operations & Algebraic Thinking						
Use the four operations with whole numbers to solve problems.	X	X				
Gain familiarity with factors and multiples.	X	X				
Generate & analyze patterns.	X	X	X	X	X	X
Number and Operations in Base Ten						
Generalize place value understanding for multi-digit whole numbers.	X	X		X		
Use place value understanding and properties of operations to perform multi-digit arithmetic.	X	X		X		
Number and Operations—Fractions						
Extend understanding of fraction equivalence and ordering.			X			
Build fractions from unit fractions by applying and extending previous understandings of operations on whole numbers.			X			
Understand decimal notation for fractions, and compare decimal fractions.				X		
Grade 5						
Operations & Algebraic Thinking						
Write and interpret numerical expressions.					X	
Analyze patterns and relationships.	X	X	X	X	X	X
Number and Operations in Base Ten						
Understand the place value system.	X	X		X	X	
Perform operations with multi-digit whole numbers and with decimals to the hundredths.	X	X		X		
Number and Operations – Fractions						
Use equivalent fractions as a strategy to add and subtract fractions.			X			
Apply and extend previous understandings of multiplication and division to multiply and divide fractions.			X			
Grade 6						
Ratios and Proportional Relationships						
Understand ratio concepts and use ratio reasoning to solve problems.			X			X
The Number System						
Apply and extend previous understandings of multiplication and division to divide fractions by fractions.			X			
Compute fluently with multi-digit numbers and find common factors and multiples.	X	X		X		

	Ch. 1: Multiplication	Ch. 2: Division	Ch. 3: Fractions	Ch. 4: Decimals	Ch. 5: Algebra / Mental Math	Ch. 6: Word Problems
Grade 6 (cont.)						
The Number System (cont.)						
Apply and extend previous understandings of numbers to the system of rational numbers.			X	X	X	
Expressions and Equations						
Apply and extend previous understandings of arithmetic to algebraic expressions.					X	X
Reason about and solve one-variable equations and inequalities.					X	X
Represent and analyze quantitative relationships between dependent and independent variations.						X
Grade 7						
Ratios and Proportional Relationships						
Analyze proportional relationships and use them to solve real-world and mathematical problems.						X
The Number System						
Apply and extend previous understandings of operations with fractions to add, subtract, multiply, and divide rational numbers.					X	
Expressions and Equations						
Use properties of operations to generate equivalent expressions.						X
Solve real-life and mathematical problems using numerical and algebraic expressions and equations.						X
Grade 8						
The Number System						
Know that there are numbers that are not rational, and approximate them by rational numbers.						
Expressions and Equations						
Work with radicals and integer exponents.					X	
Understand the connections between proportional relationships, lines, and linear equations.						
Analyze and solve linear equations and pairs of simultaneous linear equations.					X	X
Functions						
Define, evaluate, and compare functions.						X
Use functions to model relationships between quantities.					X	X

Chapter 1

Multiplication

Chapter Introduction

What Is Multiplication?

Multiplication is the inverse of division, in the same way that addition is the inverse of subtraction. In other words, multiplication has the opposite effect of division. For example, $3 \times 3 = 9$ and $9 \div 3 = 3$.

Multiplication makes repeated addition more efficient by making copies of an original number. A multiplication or times table is a simple way to see multiplication patterns. Multiplication can also be visualized by counting same-sized objects arranged in a rectangle or by finding the area of a rectangle whose sides have given lengths.

Vocabulary of Multiplication

Before introducing one or more of the reteaching models proposed in this chapter, you may need to review the following multiplication terms with your students.

In a multiplication problem the first number is the *multiplicand*, the second is the *multiplier*, and the answer is the *product*. The multiplicand and multiplier are *factors* of the product. Factors are either composite numbers (like 4), which have their own factors (2×2) or prime numbers whose only factors are itself and 1 (like 3). A *multiple* of a number is the product of that number and any other whole number. For example, 12 is a multiple of 2, 3, 4, and 6.

Properties of Multiplication

Some students may need to be reminded of the various properties of multiplication:

Multiplication has certain properties	
Commutative Property	The order in which two numbers are multiplied does not matter: $6 \times 4 = 4 \times 6$.
Associative Property	When three or more numbers are multiplied, the product is the same regardless of the grouping of the factors: $(2 \times 3) \times 4 = 2 \times (3 \times 4)$.
Distributive Property	The multiplicand or multiplier can be decomposed and multiplied individually by the multiplier and then added together: $6 \times (2 + 3) = (6 \times 2) + (6 \times 3)$.
Identity Property	Anything multiplied by 1 is itself: $27 \times 1 = 27$.
Inverse Property	Every number, except 0, has a multiplicative inverse: $\frac{1}{2} \times \frac{2}{1} = 1$.
Zero Property	Anything multiplied by 0 is 0: $72 \times 0 = 0$.

Expectations for Middle School

Multiplication is typically introduced in the second-grade math curriculum as repeated addition, and it is fully developed in third and fourth grades. By the end of grade four, ages 9–10, students are expected to be able to fluently multiply and divide within 100 and to have memorized the times tables, also called "multiplication facts." This lays the foundation for middle school students to be able to fluently add, subtract, multiply, and divide whole numbers using standard algorithms (methods) for each operation. In middle school, they are expected to begin multiplying fractions and decimals, which lays a foundation for algebra. If students are not fluent in whole number multiplication by sixth grade, they need specific help to address their difficulties.

Common Problems Students Have with Multiplication

Students learn multiplication in stages and can get stuck in any stage. Multiplication is founded on an understanding of addition. The basic stages of learning multiplication and the difficulties associated with those stages are listed following.

Multiplication as Repeated Addition Understands that multiplication is repeated addition.	*Students may not understand the relationship of multiplication to addition.*
Multiplication Table Patterns Recognizes that multiplication facts can be represented in rectangular arrays, with one factor the number of rows and the other factor the number of columns.	*Students may not recognize the patterns in a multiplication table.*

Properties of Multiplication Understands that, like addition, multiplication has the following properties:
  a. Commutative: $3 \times 6 = 6 \times 3$
  b. Associative: $(2 \times 3) \times 4 = 2 \times (3 \times 4)$
  c. Distributive: $5 \times (3 + 4) = (5 \times 3) + (5 \times 4)$
	If students don't understand the properties of multiplication, they will not be able to use multiplication effectively.

Multiplication Is Different from Addition Recognizes that multiplication is different from addition.
  a. Multiplying a number by 1 is the number.
  b. Multiplying a number by 0 is 0.
	Until students understand the concept of multiplication, they will not understand how it is different from addition.

Multiplication and Division as Inverse Operations Recognizes that multiplication and division are inverse operations. ($2 \times 3 = 6$; therefore, $6 \div 3 = 2$.)	*If students don't understand that multiplication and division are inverse operations, they will double their efforts in learning multiplication facts.*

Multiplication Operates on Equal Groups Understands that multiplication and division apply to situations with equal groups, arrays, or areas.	*Unless students can visualize multiplication and division as operations on equal groups, they will have difficulty recognizing the efficiency of these operations over addition and subtraction.*
Distributive Property in Multiplication Algorithms Understands that the distributive property is at the heart of strategies and algorithms for multiplication and division. The distributive property makes it possible to multiply 4 × 7 by decomposing 7 as 5 + 2, and 4 × 7 = 4 × (5 + 2) = 4 × 5 + 4 × 2 = 20 + 8 = 28.	*If students do not understand that they can use the distributive property for multiplication, they will not be able to use it for long division or algebraic concepts and will not gain fluency in mathematics.*
Multiplication Fact Mastery Knows single-digit multiplication facts. Multiplication facts are often taught in a specific order, as shown on page 11.	*Unless students have mastered the multiplication facts, they will have great difficulty attaining fluency.*
Alternative Methods for Solving Multiplication Problems Understands how the distributive property and the expanded form of a multi-digit number can be used to calculate products of multi-digit numbers.	*Recognizing that there is more than one way to solve a multiplication problem is a sign that students understand how multiplication works and have not simply memorized an algorithm.*
Mental Multiplication Mentally calculates products of one-digit numbers and one-digit multiples of 10, 100, 1,000, and so on (6 × 1,000 = 6,000).	*If students can calculate multiples of ten, they demonstrate understanding of place value and multiplication.*
Assessing Reasonableness of Answers Assesses the reasonableness of answers using mental computation and estimation strategies, rounding to the nearest 10, 100, 1,000, and so on.	*When students check the reasonableness of multiplication using mental computation and estimation, they demonstrate competence in place value and multiplication.*

Using the Standard Multiplication Algorithm for Two-Digit Numbers Computes the products of two-digit numbers using the standard algorithm and checks the result using estimation.	*Until students achieve fluency using the standard algorithm with two-digit numbers, they will not have developed a full understanding of and an ability to use multiplication.*
Computing Products of Three-Digit Numbers Computes the products of three-digit and higher numbers using the standard algorithm.	*When students can compute the products of higher numbers, they demonstrate fluency with whole number multiplication.*
Multiplication of Fractions as Repeated Addition Understands that multiplying fractions by whole numbers comes from repeated addition: $\frac{1}{3} + \frac{1}{3} + \frac{1}{3} = \frac{1}{3} \times \frac{3}{1} = 1$.	*Unless students recognize the relationship to repeated addition, they run the risk of mindlessly following a formula.*

What Research Says

Much less research is available on single-digit multiplication and division than on addition and subtraction. The research that does exist confirms these strategies:

- Students invent many of the procedures they use for multiplication. They find patterns and use skip counting, counting by 2s, 3s, or another multiple.

- Treating multiplication learning as pattern finding simplifies the task and uses a core mathematical idea.

- After children identify patterns, they need a lot of practice and experience to gain fluency, but research has not yet identified how students acquire the fluency or what experiences might be most efficient and effective.

Models For Reteaching Multiplication Concepts

Following are three models that teachers use to reteach concepts at different stages of understanding the concept of multiplication. The first demonstrates multiplication as a two-dimensional model. The second emphasizes memorization of the multiplication facts. And the third introduces the multiplication algorithm for problems beyond 9 × 9.

It is strongly recommended that you use Base 10 blocks or color tiles to model the concepts. Using manipulatives not only helps students understand multiplication and division instead of simply trying to memorize facts and algorithms, but it also produces enormous benefits for a better understanding of algebraic ideas.

Model:
Multiplication as a Two-Dimensional Model

Understands that multiplication and division apply to situations with equal groups, arrays, or areas, and compares those situations.

Demonstrate how addition may be thought of as a linear operation.

If we want to add 5 and 3 on a number line, we might proceed as follows:

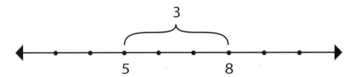

Start at 5. Count three spaces. End at 8. We're measuring in a straight line, a *linear* operation.

Demonstrate how multiplication can be reflected as sets of the same quantity.

If we multiply two quantities, however, the measure could be thought of as two-dimensional rather than linear (as in a number line). For example:

Two sets of 4 rectangles = 8 rectangles.

2 × 4 = 8

Granted, these rectangles could be arranged in a straight line and counted or added, but we want students to commit these multiplication facts to memory and thus reduce the need to count or add them.

Practicing the placement of tiles in various rectangular arrangements soon demonstrates how the different multiplication facts are derived and how some numbers (prime numbers) can't produce more than two arrangements.

Suppose you have 12 tiles. Six rectangular arrangements can be made:

$$12 \times 1 \text{ or } 1 \times 12, 3 \times 4 \text{ or } 4 \times 3, 6 \times 2 \text{ or } 2 \times 6$$

For 9 tiles, only three rectangular arrangements can be made:

$$3 \times 3, 9 \times 1, \text{ or } 1 \times 9$$

For 5 tiles (a prime number), only two rectangular arrangements can be made:

$$5 \times 1 \text{ or } 1 \times 5$$

Notice that the 9×1 or 1×9 array can be rotated to give the same design (the commutative property of multiplication). This can be stated another way: *Multiplication can be thought of as finding an area—a two-dimensional model.*

Models with one-digit numbers are simple, but as we move to more complicated areas using larger numbers, the models become more difficult, as you will see in the next section.

Demonstrate how multiplication can be seen as finding an area.

Multiplication is often said to be a shortcut to repeated addition. That is one way to look at it, but it can also be seen as finding an area. For example:

$$2 \times 8 =$$

$$2 + 2 + 2 + 2 + 2 + 2 + 2 + 2 = 16$$

or 8 sets of 2 = 16

$2 \times 8 = 16$ is a multiplication fact. A 2×8 rectangle also gives an area of 16 units.

Let's look at another example:

3 × 12 =

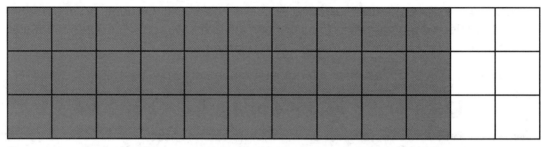

3 + 3 + 3 + 3 + 3 + 3 + 3 + 3 + 3 + 3 + 3 + 3 = 36

or 12 sets of 3

However, as illustrated by the shaded and unshaded squares above, we can also look at this as 10 sets of 3 and 2 sets of 3.

3(10 + 2) = 3(10) + 3(2) = 30 + 6 = 36

Introduce a multiplication algorithm to reflect the area example.

An *algorithm* is a systematic way of solving a problem based on patterns observed from investigations. There are several multiplication algorithms. Taking 3 × 12 = 36 as an example, sometimes we prefer to write the algorithm this way:

Example

$$
\begin{array}{r}
12 \\
\times\ 3 \\
\hline
6 \quad = 3 \times 2 \\
30 \quad = 3 \times 10 \\
\hline
36 \\
\end{array}
$$

Assess students.

Have students create a rectangular array to demonstrate a multiplication problem, such as 8 × 7 = 56.

Model:
Fact Memorization | Knows single-digit multiplication facts.

Test students to see what facts they know.

Before we examine more multiplication algorithms, let's take a closer look at the multiplication facts. Multiplication facts can be generated by examining arrays made with manipulatives (such as color tiles) or by drawing.

Notice that any way we look at the array on page 7— 2 × 8 or 8 × 2—it equals 16, a multiplication fact.

The teaching of multiplication facts is very important and a skill that should have been mastered in the early grades. Multiplication facts need to be memorized.

On the next page is a pretest that you may give to students to determine which facts they have already memorized.

1. 2 × 2 = ☐   **2.** 2 × 7 = ☐   **3.** 3 × 9 = ☐

4. 3 × 3 = ☐   **5.** 2 × 8 = ☐   **6.** 4 × 5 = ☐

7. 4 × 4 = ☐   **8.** 2 × 9 = ☐   **9.** 4 × 6 = ☐

10. 5 × 5 = ☐   **11.** 5 × 6 = ☐   **12.** 4 × 7 = ☐

13. 6 × 6 = ☐   **14.** 5 × 7 = ☐   **15.** 4 × 8 = ☐

16. 7 × 7 = ☐   **17.** 5 × 8 = ☐   **18.** 4 × 9 = ☐

19. 8 × 8 = ☐   **20.** 5 × 9 = ☐   **21.** 6 × 7 = ☐

22. 9 × 9 = ☐   **23.** 3 × 4 = ☐   **24.** 6 × 8 = ☐

25. 2 × 3 = ☐   **26.** 3 × 5 = ☐   **27.** 6 × 9 = ☐

28. 2 × 4 = ☐   **29.** 3 × 6 = ☐   **30.** 7 × 8 = ☐

31. 2 × 5 = ☐   **32.** 3 × 7 = ☐   **33.** 7 × 9 = ☐

34. 2 × 6 = ☐   **35.** 3 × 8 = ☐   **36.** 8 × 9 = ☐

Have students memorize essential facts.

Research shows that the order of memorization of multiplication facts is key to retention. The best order to use to memorize the facts, based on this research, is squares first, followed by the twos table, fives table, and finally the threes, fours, sixes, sevens, and eights table, in that order.

If you know the 36 facts in the following table, you know *all* the multiplication facts because the communicative property tells us that order does not make any difference. The identity property says that $1 \times A = A$, and the zero property says that $0 \times A = 0$. Drill the students on any facts they can't answer quickly.

The True Multiplication Facts ...

$2 \times 2 = 4$, $2 \times 3 = 6$, $2 \times 4 = 8$, $2 \times 5 = 10$, $2 \times 6 = 12$, $2 \times 7 = 14$, $2 \times 8 = 16$, $2 \times 9 = 18$

$3 \times 3 = 9$, $3 \times 4 = 12$, $3 \times 5 = 15$, $3 \times 6 = 18$, $3 \times 7 = 21$, $3 \times 8 = 24$, $3 \times 9 = 27$

$4 \times 4 = 16$, $4 \times 5 = 20$, $4 \times 6 = 24$, $4 \times 7 = 28$, $4 \times 8 = 32$, $4 \times 9 = 36$

$5 \times 5 = 25$, $5 \times 6 = 30$, $5 \times 7 = 35$, $5 \times 8 = 40$, $5 \times 9 = 45$

$6 \times 6 = 36$, $6 \times 7 = 42$, $6 \times 8 = 48$, $6 \times 9 = 54$

$7 \times 7 = 49$, $7 \times 8 = 56$, $7 \times 9 = 63$

$8 \times 8 = 64$, $8 \times 9 = 72$

$9 \times 9 = 81$

For more help with teaching memorization of the multiplication facts, refer to the book *Multiplication Facts in Seven Days*, published by Didax Educational Resources.

Have students demonstrate knowledge of multiplication facts.

Have students use Base 10 blocks or color tiles to demonstrate that they know the 36 multiplication facts (shown in the white squares in the following table). Have them notice that any way they look at a particular array—say, 2 by 8 or 8 by 2—it equals a *fact* (in this case 16).

×										
	0	0	0	0	0	0	0	0	0	0
	0	1	2	3	4	5	6	7	8	9
	0	2	4	6	8	10	12	14	16	18
	0	3	6	9	12	15	18	21	24	27
	0	4	8	12	16	20	24	28	32	36
	0	5	10	15	20	25	30	35	40	45
	0	6	12	18	24	30	36	42	48	54
	0	7	14	21	28	35	42	49	56	63
	0	8	16	24	32	40	48	56	64	72
	0	9	18	27	36	45	54	63	72	81

Model:
Long Multiplication Algorithm

Computes the products of two-digit numbers using the standard algorithm and checks the results using estimation.

Introduce the multidigit multiplication algorithm.

Any multiplication beyond 9 × 9 involves the use of an algorithm. Multiplication algorithms involve facts, but 12 × 13, for example, is not a fact. The product of 12 × 13 uses the facts 2 × 3 = 6, 3 × 1 = 3, 1 × 2 = 2, and 1 × 1 = 1. Putting these facts together in a systematic way is the algorithm.

Note that the 3 × 1 in 12 × 13 is actually 3 times 1 ten, and the 1 × 1 is actually 10 ones times 10 ones because of its place value.

First, let's visualize the product of 12 × 13 as a rectangular arrangement. This means we have a length of 12 units on one side of a rectangle and a length of 13 units on the other side. The question is, how many units are in the rectangle? True, we could simply count them or add each row or column (repeated addition), but an easier and faster method is desired.

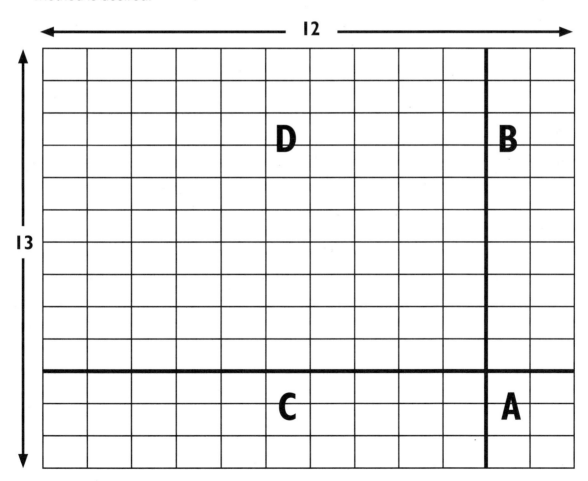

The method most frequently used is called the "long multiplication algorithm" and simply adds up the four different sections.

Example
A = 3 × 2 = 6
B = 2 × 10 = 20
C = 3 × 10 = 30
D = 10 × 10 = 100
A + B + C + D = 156
Total = 156

Assess students.

To determine if students can use the algorithm effectively, observe them as they solve the following multidigit multiplication problems with Base 10 blocks.

10 × 3   28 × 9   11 × 23   43 × 5

Chapter Wrap-Up

Practice Ideas

Fact Practice Reinforce the multiplication facts by having students practice and then test one another on the different facts beginning with the easiest and progressing to the most difficult to remember: 0, 1, squares, 2, 5, 3, 4, 6, 7, 8. Reinforce the patterns in the multiplication table.

Multiples of 10 Practice Reinforce the multiples of 10 concept by having students practice mental math multiplication by 10s, 100s, 1,000s, and so on.

For example: $2 \times 1{,}000 = 2{,}000$, $35 \times 100 = 3{,}500$, $4 \times 1{,}000{,}000 = 4{,}000{,}000$.

Inverse Operations Practice Have students provide the inverse equations for multiplication and division problems: 3×4 and $12 \div 3$, 23×3, and $69 \div 23$.

Have students practice algorithms.

Have students practice using the multiplication algorithms by challenging pairs of students to see who can calculate the problems more quickly.

$$\begin{array}{r} 34 \\ \times\ 13 \\ \hline \end{array} \qquad \begin{array}{r} 263 \\ \times\ 24 \\ \hline \end{array}$$

Further Reteaching

The teaching models in this chapter demonstrate how teachers can use manipulatives and visual models to reteach multiplication concepts. Because multiplication is foundational and whole number multiplication is concrete and not abstract, using manipulatives to reteach multiplication concepts is a very effective method for struggling students to visualize the multiplication facts and patterns. Having students demonstrate their understandings with manipulatives is an excellent way to confirm that they have internalized the concepts and are not just trying to memorize and guess at answers without understanding.

Skill Maintenance and Assessment

Use the following assessments to assess and refresh student understanding of the multiplication facts.

Multiplication Facts

1. 7 × 6 = ☐   **2.** 9 × 2 = ☐   **3.** 3 × 3 = ☐

4. 2 × 3 = ☐   **5.** 6 × 6 = ☐   **6.** 4 × 2 = ☐

7. 3 × 7 = ☐   **8.** 7 × 5 = ☐   **9.** 5 × 3 = ☐

10. 5 × 2 = ☐   **11.** 8 × 6 = ☐   **12.** 6 × 3 = ☐

13. 4 × 3 = ☐   **14.** 4 × 8 = ☐   **15.** 8 × 2 = ☐

16. 2 × 4 = ☐   **17.** 9 × 8 = ☐   **18.** 6 × 8 = ☐

19. 3 × 2 = ☐   **20.** 9 × 9 = ☐   **21.** 9 × 4 = ☐

22. 6 × 9 = ☐   **23.** 5 × 8 = ☐   **24.** 2 × 2 = ☐

25. 7 × 4 = ☐   **26.** 7 × 9 = ☐   **27.** 7 × 3 = ☐

28. 8 × 5 = ☐   **29.** 8 × 6 = ☐   **30.** 9 × 6 = ☐

31. 2 × 5 = ☐   **32.** 7 × 7 = ☐   **33.** 5 × 5 = ☐

34. 3 × 4 = ☐   **35.** 4 × 9 = ☐   **36.** 3 × 5 = ☐

37. 8 × 8 = ☐   **38.** 7 × 9 = ☐   **39.** 4 × 6 = ☐

40. 3 × 8 = ☐   **41.** 9 × 4 = ☐   **42.** 4 × 7 = ☐

43. 3 × 9 =

44. 8 × 3 =

45. 6 × 4 =

46. 8 × 7 =

47. 7 × 8 =

48. 8 × 4 =

49. 7 × 6 =

50. 2 × 8 =

51. 4 × 4 =

52. 6 × 5 =

53. 6 × 2 =

54. 9 × 3 =

55. 8 × 7 =

56. 9 × 5 =

57. 2 × 9 =

58. 9 × 7 =

59. 7 × 6 =

60. 9 × 6 =

61. 7 × 5 =

62. 9 × 7 =

63. 9 × 7 =

64. 9 × 6 =

65. 2 × 7 =

66. 7 × 9 =

67. 4 × 9 =

68. 5 × 6 =

69. 5 × 7 =

70. 6 × 9 =

71. 5 × 4 =

72. 4 × 7 =

73. 6 × 7 =

74. 7 × 2 =

75. 8 × 9 =

76. 6 × 8 =

77. 3 × 6 =

78. 7 × 6 =

79. 8 × 7 =

80. 2 × 6 =

81. 9 × 8 =

82. 9 × 3 =

83. 4 × 5 =

84. 5 × 9 =

Chapter 2

Division

Chapter Introduction

What Is Division?

Division is the inverse of multiplication, in the same way that subtraction is the inverse of addition. Because division and multiplication are inverse operations, once the multiplication facts are learned, the division facts are learned, too. The multiplication or times table is the same as the division table and is a simple way to see multiplication and division patterns. Just as rectangular arrays can aid in the understanding of multiplication, division problems can be visualized by counting same-sized objects arranged in a rectangle or by finding the side of a rectangle when the other side and the area are known.

Unlike multiplication, however, division often results in remainders. Fractions and decimals are calculated to complete the division of a remainder.

Before introducing the reteaching models proposed in this chapter, you may want to review the vocabulary and properties of division, as described following, for those students who need it.

Vocabulary of Division

In a division problem ($8 \div 2 = 4$) the number being divided is the *dividend*, the number divided by is the *divisor*, and the answer is the *quotient*.

Fractions represent division problems, as well. The top number (the *numerator*) is the dividend and the bottom number (the *denominator*) is the divisor. Beyond elementary school, division is typically presented as a fraction.

Properties of Division

Division has certain properties.	
Distributive Property Over Addition	The Distributive Property makes long division possible. It is the property used when you regroup tens and ones so that the divisor can be divided into the dividend. This property allows for the dividend or divisor to be "decomposed" into its factors and divided individually by the divisor and then added together: $84 \div 4 = (80 \div 4) + (4 \div 4) = 20 + 1 = 21$.
Identity Property	Anything divided by 1 is itself: $27 \div 1 = 27$.
A number divided into itself is equal to 1: $8 \div 8 = 1$.	
Division by 0 is undefined. Although a number multiplied by 0 is 0, a number cannot be divided by 0. Calculators will display an error message when a number is divided by 0.	

Expectations for Middle School

Division typically is introduced in the second- and third-grade math curriculum in sharing problems, and it is fully developed in fourth and fifth grades. By the end of grade five, ages 9–11, students are expected to be able to fluently multiply and divide within 100 and have memorized the times tables, also called *facts*. At this age they should be able to complete division problems with and without remainders. This lays the foundation for middle school students to be able to fluently add, subtract, multiply, and divide whole numbers using standard *algorithms* (methods) for each operation. In middle school, they are expected to begin multiplying and dividing fractions and decimals, which lays a foundation for algebra. If students are not fluent in whole number division by sixth grade, they need specific help to address their difficulties.

Common Problems Students Have with Division

Students learn division in stages, and they can get stuck in any stage. Division is founded on an understanding of multiplication, which is founded on an understanding of addition. The basic stages of learning division and the difficulties associated with those stages are listed following.

Division as the Inverse of Multiplication Understands that division is the opposite of multiplication.	*When students don't understand the relationship between multiplication and division, they are doubling their efforts in learning facts and will not be able to use multiplication and division fluently.*
Division in Arrays Visualizes multiplication and division facts in arrays to reinforce the idea of multiplying and dividing equal groups, which is the efficiency of multiplication and division over addition and subtraction.	*When students don't recognize the relationship between multiplication and division, they may try to memorize facts individually and will not be able to recognize and use number patterns and relationships.*

Division Is Different from Multiplication Recognizes that, unlike multiplication, division does not have the Associative or Commutative Properties. It does matter in division which number is the divisor and which is the dividend. Division can also result in remainders that need to be addressed. Sometimes remainders can be expressed as fractions or decimals, but other times—for example, if you are dividing children into groups—it doesn't make sense to have a fraction.	*Unless students realize the significant differences between multiplication and division, they can make serious errors and will not understand the concept of division.*
Division Operates on Equal Groups Recognizes that multiplication and division apply to increasing and decreasing equal groups. Addition and subtraction apply to individual things. This makes multiplication and division very efficient because you don't have to add and subtract groups over and over. You can simply multiply or divide.	*When students focus on mathematical procedures and don't understand the underlying concepts, they compute meaninglessly and may calculate results that make no sense.*
Distributive Property over Addition Uses the Distributive Property of regrouping numbers to divide them more efficiently.	*Unless students understand how numbers can be regrouped and recombined and still maintain the quantity, they will not be able to fluently divide larger digits into smaller digits.*
Multiples of 10 in Division Estimates and checks the reasonableness of answers by mentally calculating quotients of one-digit numbers and one-digit multiples of 10, 100, 1,000, and so on (6,000 ÷ 1,000 = 6).	*If students cannot mentally multiply and divide multiples of ten, they do not understand place value, which can result in significant mistakes in calculations.*
Reasonableness of Quotients Assesses the reasonableness of answers using mental computation and estimation strategies. Rounding to the nearest 10, 100, 1,000, and so on is a double check on any calculation.	*If students cannot check the reasonableness of division answers using mental computation and estimation, they lack competence in place value understanding and division.*

Division With and Without Remainders Computes the quotients of two- and three-digit numbers using the standard algorithm with and without remainders, and checks the result using estimation.	*If students do not address remainders reasonably, they do not demonstrate understanding of division.*
Standard Algorithm in Long Division Computes the quotients of four-digit and higher dividends using the standard algorithm.	*If students cannot use the long division algorithm with greater numbers, they cannot develop full understanding of and fluency with division.*
Fractions as Division Understands that fractions give meaning to the quotient of any whole number. $\frac{3}{4}$ is $3 \div 4$.	*Unless students recognize that a fraction represents division of a part into a whole, they will not develop a foundation for algebra or have a true understanding of fractions.*
Dividing Decimals Understands how to divide decimals and where to place the decimal point.	*If students do not recognize where to place a decimal in a division problem, they do not understand the concept of decimal fractions.*
Dividing Fractions by Whole Numbers Understands that dividing a unit fraction by a whole number results in a smaller unit fraction. $\frac{1}{2} \div 4 = \frac{1}{8}$.	*Until students can explain the counterintuitive concept of division of fractions by whole numbers, they do not fully understand division.*
Dividing Whole Numbers by Fractions Understands that dividing a whole number by a unit fraction results in a greater whole number. $8 \div \frac{1}{2} = 16$.	*Unless students can explain this counterintuitive concept, they will not recognize the differences in working with fractions and whole numbers.*

What Research Says

Much less research is available on single-digit multiplication and division than on addition and subtraction. The research that does exist confirms these strategies:

- Children invent many of the procedures they use for division. They find patterns and use skip counting backwards, counting by 2s, 3s, or another multiple.

- Treating division learning as pattern finding simplifies the task and uses a core mathematical idea.

- After children identify division patterns, they need a lot of practice and experience to gain fluency, but research has not yet identified how children acquire the fluency or what experiences might be most efficient and effective.

Models For Reteaching Division Concepts

Following are two models that teachers can use to reteach division concepts. The first shows how to introduce division, relating it to multiplication through visual models. The second presents the thought processes and skills involved in long division: rounding, estimating, multiplying by multiples of 10, and compatible numbers.

To model division for struggling students, use Base 10 blocks or color tiles whenever it makes sense. Using manipulatives not only helps students understand multiplication and division instead of simply trying to memorize facts and algorithms, but it also produces enormous benefits for a better understanding of algebraic ideas.

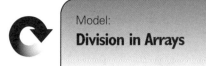

Model:
Division in Arrays

Visualize division facts in arrays to reinforce the ideas of division as the inverse of multiplication and of dividing *equal* groups, which is the efficiency of division over subtraction.

Use Base 10 blocks to model a division problem.

Because multiplication and division are inverse operations, division is simply the process of finding a missing factor from a multiplication problem. Say you want to find the answer to 143 ÷ 13. You are also looking for 13 × ? = 143.

Base 10 blocks are a wonderful way to model division. Taking the preceding example, 143 can be modeled with one Hundred flat, four Ten rods, and three Unit cubes. You want to arrange these units into a rectangle so that the dimension of one side is 13 units long. The figure below illustrates the solution.

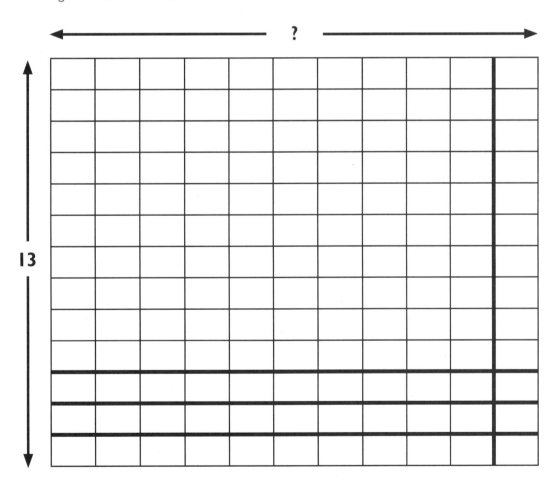

Now let's count to find the measurement of the side in question. By counting the units on that side of the rectangle, you'll see the number is 11. So 143 ÷ 13 = 11. We can check this by multiplying 11 × 13 and finding that we get 143.

Related Equations	
13 × 11 = 143	143 ÷ 13 = 11
11 × 13 = 143	143 ÷ 11 = 13

Use Base 10 blocks to model the division algorithm.

Now, let's see how 143 ÷ 13 could be solved without using Base 10 blocks. Let's look again at what the array tells us:

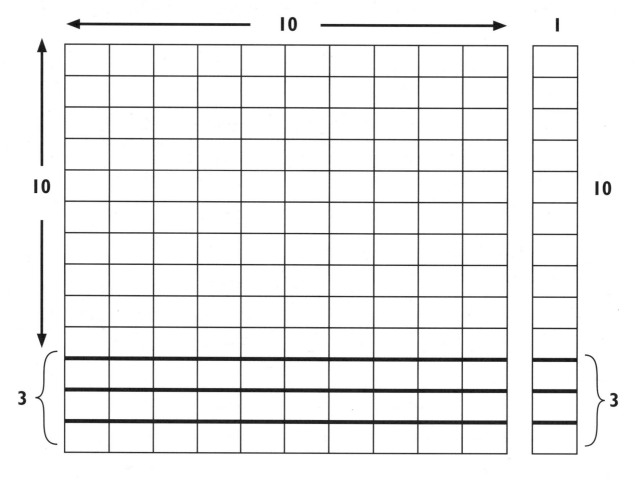

First, we can see that there are really two smaller rectangles contained in the whole rectangle: a 10 × 13 rectangle and a 1 × 13 rectangle. The 10 × 13 rectangle comprises 130 units. 10 × 13 = 130. The 1 × 13 rectangle comprises 13 units. So the sum of the two rectangles is 143 units. Separating 130 from 143 demonstrates the first calculation in the following division algorithm.

Introduce the division algorithm.

Once you see what the numbers represent with Base 10 blocks, you can illustrate the problem of 143 ÷ 13 this way.

$$
\begin{array}{r}
10 + 1 \\
13\overline{)\,143} \\
\underline{130} \\
13 \\
\underline{13} \\
0
\end{array}
$$

A shorter version of the problem can be illustrated by simply placing the one ten in the tens place above the 143. The shorter version is what we generally teach as the traditional division algorithm.

$$
\begin{array}{r}
11 \\
13\overline{)\,143} \\
\underline{130} \\
13 \\
\underline{13} \\
0
\end{array}
$$

Assess students.

Have students demonstrate these division problems using Base 10 blocks and then represent them using the division algorithm.

1. $10\overline{)100}$ 2. $24\overline{)192}$ 3. $15\overline{)135}$

4. $14\overline{)196}$ 5. $18\overline{)162}$

Model:

Standard Algorithm in Long Division

Computes the quotients of four-digit and higher dividends using the standard algorithm.

Introduce the long division algorithm.

An *algorithm* is a systematic method for solving a math problem. The long division algorithm involves several steps and several different skills.

The recommended sequence for teaching the long division algorithm is:

1. rounding numbers,

2. using compatible numbers,

3. multiplying by multiples of 10,

4. estimating, and

5. dividing.

The following sections demonstrate each skill and then show how all the steps are used in the long division process. Even though calculators can be used to solve long division problems quickly, it is important for students to understand the thought processes involved in long division so they can check the reasonableness of their answers.

Introduce rounding.

There are several methods of rounding, but they really mean only one thing: what is the number closest to? First, we need to know closest to what? Tens, hundreds, thousands, tenths, hundredths? Once that is decided, the process is easy. If the number to be rounded is 5 or more in the "rounding place," round up. If less than 5, round down.

For example:

2,365 rounded to the nearest 10 is 2,370

2,365 rounded to the nearest 100 is 2,400

2,365 rounded to the nearest 1,000 is 2,000

Here are a few more:

5,672 rounded to the nearest 10 is 5,670

5,672 rounded to the nearest 100 is 5,700

5,672 rounded to the nearest 1,000 is 6,000

47.385 rounded to the nearest one is 47

47.385 rounded to the nearest tenth is 47.4

47.385 rounded to the nearest hundredth is 47.39

Have students practice rounding.

Practice
Try rounding these numbers: **1.** 1,234 to the nearest 100 **2.** 3,456 to the nearest 10 **3.** 7,895 to the nearest 1,000 **4.** 32.7 to the nearest whole number **5.** 39.68 to the nearest 10 **6.** 12.35 to the nearest tenth

Introduce compatible numbers.

One way to compute long division is to use compatible numbers to estimate and check the reasonableness of an answer. Compatible numbers are numbers that "go together" easily and are easy to calculate in your head, such as 30 and 50, 100 and 200, 1,200 and 1,400, and, for some people, numbers ending in 5, like 15 and 25. Compatible numbers are especially useful when finding estimates or performing mental math.

Here are a few examples:

Estimate the sum of 396 and 432. Two compatible numbers may be 400 and 400 or 400 and 430. Either will give a fairly good estimate of 800 or 830, respectively. (828 is the exact answer.)

Estimate the product of 26 × 32. Two compatible numbers may be 30 and 30, whose product is 900. Or choose 25 × 30, whose product is 750. Either may be a reasonable estimate, depending on how the product is to be used. The exact answer is 832.

There will be times when you may want to find two estimates and select something in between. Looking at 26 × 32 again, 30 × 30 = 900 and 25 × 30 = 750. The difference is 150, so you could estimate the answer this way: $\frac{1}{2}$ (150) + 750 = 825. This is a very close estimate; the exact answer is 832. You may want to do this when the two compatible estimates are significantly apart. However, any one of these methods of estimation could be useful.

Have students practice using compatible numbers.

Practice
Estimate the answers using compatible numbers.

1. 1,346 + 2,357 =   2. 4,639 + 2,137 =

3. 53 × 37 =   4. 280 × 43 =

5. 1,634 × 51 =   6. 140 × 160 =

7. 325 × 21 =   8. 308 × 99 =

Introduce multiplication by multiples of ten.

Another operation we need for long division is multiplication by multiples of ten. Many people can do this in their heads.

$$10 \times 10 = 100$$
$$100 \times 10 = 1{,}000$$
$$100 \times 100 = 10{,}000$$
$$100 \times 1{,}000 = 100{,}000$$
$$1{,}000 \times 1{,}000 = 1{,}000{,}000$$
$$1{,}000 \times 10 = 10{,}000$$
$$1{,}000 \times 00 = 100{,}000$$

Do you see the pattern?

$$10 \times 10 = \text{a 1 with two zeros, or } 100$$
$$100 \times 100 = \text{a 1 with four zeros, or } 10{,}000$$

Yes, just count the zeros and write 1 followed by that many zeros.

Some people like to use *scientific notation*, which is a system of writing numbers with a base of 10 and signifying the number of zeros with an exponent. The *exponent* is a mathematical way of telling how many times the base is to be multiplied by itself.

For example:

In 10^3, the ten is the *base*, and the small three is the exponent. This means ten times itself 3 times, or $10 \times 10 \times 10 = 1,000$.

$$100 \times 100 = 10^2 \times 10^2 = 10,000 \text{ or } 10^4$$

$$1,000 \times 100 = (10 \times 10 \times 10) \times (10 \times 10) = 10^3 \times 10^2 = 10^5$$

Have students practice multiplying by multiples of 10.

Practice
Solve these problems in your head, but write the answers.

    1. $100 \times 10 =$

    2. $1,000 \times 10 =$

    3. $100,000 \times 100 =$

    4. $10,000 \times 1000 =$

    5. $1,000,000 \times 10 =$

Introduce estimating.

Estimating is also an important element of long division. *Estimating* means "to guess." But not just any guess—a reasonable, "educated" guess. Here are some examples.

If you want to add 379 and 246, you might think, "379 is close to 400, and 246 is close to 200, so the answer will be close to 600." Not bad, since the answer is 625.

If you want to multiply 36 × 53, you might think, "36 is about 40, and 53 is about 50, so 40 × 50 = 2,000. The answer will be close to 2,000." Pretty close, because 35 × 53 = 1,908.

Each of us will probably develop our own system of estimation, but generally it involves rounding.

Have students practice estimating.

Practice
Estimate the answers to these problems in your head and then write your estimate.

    1. 258 + 243 =

    2. 3,007 + 2,153 =

    3. 57 × 36 =

    4. 727 + 139 =

    5. 296 ÷ 72 =

Demonstrate using these skills in long division.

Rounding, using compatible numbers, multiplying by multiples of ten, and estimating, along with division itself, are all involved in solving long division problems. The following examples and exercises introduce students to a division algorithm that emphasizes place value and uses expanded notation so students are mindful that they are multiplying multidigit numbers. This "nontraditional" method will add understanding and skill to learning the traditional algorithm. Once students have mastered this new division procedure, you may want to focus on the more traditional method for division.

- **Use rounding, estimating, multiplying by multiples of 10, and compatible numbers.**

    Suppose we want to divide 164 by 8.

    $$8\overline{)164}$$

    Let's start by *rounding* 164 to 160, because we know 160 ÷ 8 = 20—a good estimate.

    Now let's *multiply by powers of 10* (8 × 20 = 160), and do the following:

    $$
    \begin{array}{r}
    20 \\
    8\overline{)164} \\
    160 \\
    \hline
    4
    \end{array}
    $$

    That is, we multiplied 8 × 20 and got 160. The numbers 8 and 2 are compatible because when we learned our doubles facts, we learned that 8 × 2 = 16. If 8 × 2 = 16 and we know how to multiply by multiples of 10, we know that 8 × 20 = 160.

- **Deal with remainders.**

    Finally, we subtracted 160 from 164, giving us a remainder of 4. We could write the answer as "20 remainder 4" or "$20\frac{1}{2}$" because $\frac{4}{8} = \frac{1}{2}$.

- **Practice long division.**

Demonstrate this one: 347 ÷ 9.

Think:

  9 rounds to 10
  347 rounds to 300
  $10 \times ? = 300$, $10 \times 30 = 300$
  So we know there are at least 30 nines
  in 347. $9 \times 30 = 270$. Record it.

$$
\begin{array}{r}
30 \\
9\overline{)347} \\
270 \\
\hline
77 \\
63 \\
\hline
14 \\
9 \\
\hline
5
\end{array}
$$

Think:

  How many 10's in 70? Answer: 7.
  So $7 \times 9 = 63$ remainder 14.
  How many 9's are in 14? Answer: 1.
  Remainder 5.

Answer: 38 r. 5

Have students practice long division.

Practice

Solve.

1. $7\overline{)293}$ 2. $8\overline{)964}$ 3. $9\overline{)1{,}243}$

4. $6\overline{)992}$ 5. $7\overline{)463}$ 6. $6\overline{)520}$

7. $9\overline{)1{,}309}$ 8. $8\overline{)931}$ 9. $5\overline{)531}$

10. $7\overline{)364}$ 11. $12\overline{)1{,}320}$ 12. $13\overline{)1{,}492}$

13. $17\overline{)3{,}096}$

Develop students' expertise in dividing with two-digit divisors less than 20.

Start by demonstrating 158 ÷ 12.

Let's round 12 to 10, a nice, easy, compatible number, and 158 to 160, also a nice, compatible number. 160 ÷ 10 = 16. This is an estimate and a good place to start. But we already know that 160 is more than 158, so 16 is too big.

Let's try 14 next.

$$\begin{array}{r} 12 \\ \times\ 14 \\ \hline 48 \\ 12 \\ \hline 168 \end{array}$$

Hmm, 168 is more than 158, so 14 is too big, too.

So let's use 13 instead.

$$\begin{array}{r} 12 \\ \times\ 13 \\ \hline 36 \\ 12 \\ \hline 156 \end{array}$$

12 × 13 = 156. That's it!

$$\begin{array}{r} 3 \\ 10 \\ \hline 12\,\overline{)\,158} \\ 120 \\ \hline 38 \\ 36 \\ \hline 2 \end{array}$$ } 13

2   the remainder

Answer: 158 ÷ 12 = 13. r. 2

Have students practice division with two-digit divisors less than 20.

Practice		
1. $16\,\overline{)\,2{,}437}$	**2.** $19\,\overline{)\,3{,}459}$	
3. $12\,\overline{)\,1{,}658}$	**4.** $13\,\overline{)\,9{,}321}$	

Develop students' expertise in dividing with two-digit divisors greater than 20.

Dividing numbers by divisors greater than 20 is just like dividing numbers with divisors less than 20. It's a process of rounding, using compatible numbers, estimating, and checking the reasonableness of your answer.

Have students look at this example: $9,364 \div 51$.

Round compatible numbers and estimate.

$$51 \overline{)9,364}$$

Think: 51 is approximately equal to 50. (Rounding)

9,364 is approximately equal to 10,000. (Compatible numbers)

$10,000 \div 50 = 200$, a good estimate. (Estimating)

But we can see that 10,000 is more than 9,364, so 200 will be too much. Let's try 150×51.

$$
\begin{array}{r}
150 \\
\times\ 51 \\
\hline
150 \\
750 \\
\hline
7,650
\end{array}
$$

Yes, 7,650 is less than 9,364, so that will work.

$$
\begin{array}{r}
150 \\
51 \overline{)9,364} \\
7,650 \\
\hline
1,714
\end{array}
$$

When we subtract 7,650 from 9,364, we get 1,714. Now think, how many times will 51 go into 1,714? Using compatible numbers, we know that $50 \times 30 = 1,500$, so $51 \times 30 = 1,530$, and that will be less than 1,714.

Now subtract 1,530 from 1,714 and think how many times 51 will go into 184. Since 51 × 3 = 153, and that is less than 184, 3 will work. Is there a remainder?

$$
\left.\begin{array}{r}
3 \\
30 \\
150
\end{array}\right\} 183
$$

$$
\begin{array}{r}
51\overline{)9,364} \\
7,650 \\
\hline
1,714 \\
1,530 \\
\hline
184 \qquad \text{and } 51 \times 3 = 153 \\
153 \\
\hline
31 \qquad \text{the remainder}
\end{array}
$$

Answer: 183 r. 31 or $183\frac{31}{51}$

Here's another problem to try with a divisor greater than 20.

$38\overline{)1,492}$ Think: Round 38 to 40.

$40 \times 30 = 1,200$

$40 \times 40 = 1,600$

We can see that 40 is too high. So let's begin with 38 × 35. 38 × 35 = 1,330, leaving a remainder of 162.

$$
\left[\begin{array}{r}
1,492 \\
\times\ 1,330 \\
\hline
162
\end{array}\right]
$$

How many 38s or 40s are in 162? Approximately 4.

So let's try: 4 × 38 = 152. That works.

$$
\left.\begin{array}{r}
4 \\
35
\end{array}\right\} 39
$$

$$
\begin{array}{r}
38\overline{)1,492} \\
1,330 \\
\hline
162 \\
152 \\
\hline
10
\end{array}
$$

Answer: 39 r.10 or $39\frac{10}{38}$ or $39\frac{5}{19}$

Have students practice long division.

Practice
Solve these problems using rounding, compatible numbers, estimating, and checking the reasonableness of your answers. Show your work.

1. $63\overline{)1{,}937}$ **2.** $73\overline{)2{,}349}$ **3.** $64\overline{)7{,}360}$

4. $52\overline{)2{,}347}$ **5.** $49\overline{)3{,}332}$ **6.** $138\overline{)19{,}236}$

Chapter Wrap-Up

Extra Practice Ideas

Estimation with Rounding to Compatible Numbers Practice Give students 30 seconds to estimate the answer to each of the following problems, and then compare and discuss their answers and thought processes to see how rounding and compatible numbers were used to get a reasonable answer.

1. $18\overline{)398}$

2. $24\overline{)5{,}205}$

3. $22\overline{)6{,}804}$

4. $18\overline{)5{,}455}$

5. $52\overline{)72{,}125}$

Inverse Operations Practice Have students provide the inverse equations for the following expressions:

1. 3×4 2. 12×7 3. 24×25 4. 16×16

Algorithm Practice Have students practice using the long division algorithms by challenging pairs of students to see who can calculate the problems more quickly. Have students calculate any remainders as decimals.

1. $24 \overline{)263}$ 2. $22 \overline{)585}$

3. $220 \overline{)747}$ 4. $499 \overline{)1,274}$

5. $1,760 \overline{)5,280}$

Further Reteaching

- **Manipulatives** The teaching models in this chapter demonstrate how teachers can use manipulatives, visual models, and thought practices to reteach division concepts. Because division is foundational and whole number division is concrete and not abstract, using manipulatives to reteach division concepts is a very effective method for struggling students to visualize division patterns. Having students demonstrate their understandings with manipulatives is an excellent way to confirm that they have internalized the concepts and are not just trying to memorize and guess at answers without understanding.

- **Think-Aloud Process** If students continue to struggle, demonstrating the thought processes involved in long division (rounding, estimating, using compatible numbers, and multiplying by multiples of 10) will reveal that problem solving, not memorization, is the key to finding answers. Once you model thinking, have students talk out loud as they solve a problem so you can identify what is giving them trouble.

Chapter 3

Fractions

Chapter Introduction

What Are Fractions?

The word *fraction* comes from the Latin word *fractus*, which means "broken." Fractions are used to name part or equal parts of a whole object or to compare two quantities, as in a ratio. There are several ways to understand fractions, and you may want to remind your students of them.

- As the relationship of a part or parts to a whole, the top number in a fraction is the *numerator*, which represents the number of equal parts, and the bottom number is the *denominator*, which tells how many of those parts make up the whole. (If there are 8 slices of pizza and you eat 2, you have eaten $\frac{2}{8}$ of the pizza.)

- On a number line, fractions are the numbers that come between the integers or whole numbers. For example, $\frac{2}{3}$ is between 0 and 1 on the number line.

- Fractions are rational numbers—that is, they represent ratios. The fraction $\frac{2}{3}$ represents the ratio 2:3, which is also expressed as "2 of 3" or "2 to 3," as in "2 out of 3 people prefer...."

- Fractions can be expressed as division. The fraction $\frac{2}{3}$ represents $2 \div 3$.

- Decimals are fractions in which the denominator is a power of 10: $0.2 = \frac{2}{10}$ and $0.02 = \frac{2}{100}$.

- Percents are fractions in which the denominator is always 100: $33\% = \frac{33}{100}$.

- Fractions represent the remainder in a division problem. For example, $8 \div 3 = 2\frac{2}{3}$.

Vocabulary of Fractions

- *Reciprocal fractions* (also called the *multiplicative inverse)* are a pair of fractions in which the numerator and denominator are reversed. $\frac{1}{2}$ and $\frac{2}{1}$ are reciprocal fractions.

- A *proper fraction* is a fraction in which the value of the numerator is less than the denominator: $\frac{3}{4}$.

- An *improper fraction* is a fraction in which the value of the numerator is greater than or equal to the denominator: $\frac{4}{3}$.

- *Equivalent fractions* are fractions that have different numerators and denominators but name the same number. For example, $\frac{2}{3}$ and $\frac{6}{9}$ are equivalent fractions.

- A *unit fraction* is a fraction in which the numerator is 1, as in $\frac{1}{3}$.

- A *common denominator* is any nonzero number that is a multiple of the denominators of two or more fractions. A common denominator for $\frac{2}{3}$ and $\frac{3}{4}$ is 12, since the equivalent fraction for $\frac{2}{3}$ is $\frac{4}{12}$ and the equivalent fraction for $\frac{3}{4}$ is $\frac{9}{12}$. Finding a common denominator is necessary for adding or subtracting fractions.

- A *mixed number* is the sum of a whole number and a proper fraction: $2\frac{3}{4}$.

- *Lowest* or *simplest terms* describe a fraction in which the numerator and denominator have no factors in common other than 1: $\frac{1}{3}$ is a fraction in its simplest terms.

Operations with fractions have certain properties.

Operations with fractions have the same properties as whole numbers (the Associative, Commutative, Distributive, Identity, and Zero Properties). In addition, operations with fractions have these properties:

- Inverse Property: Every number except 0 has a multiplicative inverse, also called a reciprocal fraction: $\frac{1}{2} \times \frac{2}{1} = 1$.

- Any fraction in which the numerator and denominator are the same value is equal to 1: $\frac{9}{9} = 1$.

Some properties of whole numbers do not apply to fractions. For example, the two numbers that compose a common fraction (numerator and denominator) are related through multiplication and division, not addition. Therefore:

- When adding or subtracting fractions, one cannot add or subtract the denominators.

Expectations for Middle School

Fractions are introduced as early as kindergarten in discussions of sharing parts of a whole. Fractions are reinforced in the early grades in lessons on time (fractions of an hour), money (fractions of a dollar), and measurement ($\frac{1}{2}$ foot, $\frac{1}{2}$ mile). Adding and subtracting fractions is often introduced in grade 4 and may cause significant problems for students who attempt to apply their whole number understanding to the addition and subtraction, and later multiplication and division, of fractions.

By sixth grade, students should have a clear understanding of what fractions are, how they are different from whole numbers, and how they relate to decimals, percents, ratio, and proportion. They should be fluent in adding and subtracting fractions. In sixth grade, students are expected to develop fluency in multiplying and dividing fractions, which lays the foundation for algebra.

Common Problems Students Have with Fractions

Students learn about fractions in stages, and they can get stuck in any stage. The basic stages of learning about fractions and the difficulties associated with those stages are listed following.

Creating Fractions Creates equal-sized fractional parts of a whole using physical models. At this stage, a student may label parts with written fraction notation.	*Unless students can use physical models of fractions, they do not understand parts and wholes.*
Fractions as Numbers Recognizes that a fraction is a number that references a part (numerator) of a whole (denominator).	*Without this recognition, students are unable to add, subtract, multiply, or divide fractions.*
Denominator Value Recognizes that denominators represent different values from whole numbers ($\frac{1}{3}$, $\frac{1}{30}$, $\frac{1}{300}$, $\frac{1}{3,000}$).	*Unless students recognize that $\frac{1}{3,000}$ is less than $\frac{1}{3}$ and $\frac{1}{3}$ is less than 3, they do not understand the concept of fractions and will be unable to add and subtract fractions.*
Equivalent Fractions Recognizes that two fractions, such as $\frac{1}{3}$ and $\frac{2}{6}$, are equal when they represent the same portion of a whole.	*Unless students understand equivalent fractions and how to simplify fractions, they cannot find common denominators to add and subtract fractions.*
Simplifying/Reducing Fractions Simplifies or reduces fractions so that the numerator and denominator have no common factors other than 1: $\frac{6}{12} = \frac{3}{6} = \frac{1}{2}$.	*Unless students can simplify fractions, they will not be able to identify equivalent fractions or understand equivalency, which is needed for algebraic concepts.*
Ordering Positive Fractions Orders positive fractions along with whole numbers on a number line.	*Unless students recognize that a fraction like $\frac{2}{3}$ falls two-thirds of the way between 0 and 1, they do not understand how the numerator and denominator affect the value of fractions.*
Multiplying Fractions Computes the product of two fractions by multiplying numerators and denominators: $\frac{3}{4} \times \frac{1}{2} = \frac{3}{8}$.	*Until multiplying fractions makes sense to students, they will run the risk of mindlessly following a formula.*

Adding and Subtracting Same Denominator Fractions Understands that fractions with the same denominator can be added or subtracted by adding or subtracting the numerators.	*Many students attempt to generalize whole-number addition by adding both numerators and denominators.*
Adding and Subtracting Fractions by Finding a Common Denominator Adds fractions with unlike denominators by finding a common denominator.	*Until students understand common denominators and are able to calculate them, they will not be able to add or subtract fractions, order fractions, find equivalent fractions, or develop algebraic concepts.*
Regrouping Fractions for Addition and Subtraction Adds and subtracts mixed fractions by finding common denominators and regrouping: $2\frac{1}{3} - 1\frac{1}{2} = 2\frac{2}{6} - 1\frac{3}{6} = 1\frac{8}{6} - 1\frac{3}{6} = \frac{5}{6}$.	*Until students can add and subtract mixed fractions, they will not be ready to deal with algebraic concepts.*
Dividing Fractions Divides fractions using multiplication by the inverse: $\frac{1}{2} \div \frac{1}{4} = \frac{1}{2} \times \frac{4}{1} = \frac{4}{2} = 2$.	*Until students can model this operation, they may mindlessly follow the formula "Ours is not to wonder way, just invert and multiply."*
Percents and Decimals as Fractions Finds percent and decimal equivalents for common fractions.	*Until students can convert among fractions, percents, and decimals, they will not completely understand the concept of fractions.*
Ratios as Fractions Recognizes that a ratio can be expressed as a fraction and simplified, and finds equivalent ratios in the form of fractions: 3 to 1 = 3:1 = $\frac{3}{1}$.	*Unless students understand the relationship between ratios and fractions, they will not be able to work meaningfully with proportion, scale, and data and statistics.*
Mixed Number Equivalence Recognizes that a mixed number such as $3\frac{2}{5}$ represents the sum of a whole number and a fraction less than one.	*Unless students can work flexibly with whole numbers and fractions, they will not be able to understand algebraic concepts.*

What Research Says

Research has identified the following issues relating to the difficulty of learning fractions.[1]

- Students do not think of fractions as numbers that have places on the number line. Everyday encounters with fractions (store discounts, batting percentages, half a sandwich) are seen in contexts not associated with adding, subtracting, multiplying, and dividing in the whole-number system. Many students do not connect their understanding of a fraction as a thing ($\frac{1}{4}$ tank of gas) to the number.

- Students continue to use properties they learned from operating with whole numbers, even though many do not apply to fractions. To many, $\frac{2}{3}$ looks like one whole number over another. They may think that $\frac{1}{8}$ is larger than $\frac{1}{4}$ because 8 is larger than 4 or that $\frac{3}{4} = \frac{4}{5}$ because the difference between the numerator and denominator is 1 in both fractions.

- Fractions are represented in multiple ways (as fractions, decimals, and percents), and they are used in many ways (as parts of wholes, ratios, and quotients). This causes considerable confusion. Each form has many common uses in daily life that do not seem connected. The majority of students learn rules for translating between forms but understand very little about what quantities the symbols represent and make frequent and nonsensical errors.

- The numerator and denominator of a fraction are related through multiplication and division, not addition and subtraction, which causes misunderstanding when students are introduced to fractions.

- Students are less likely to have out-of-school experiences with fractions than with whole numbers, which adds to the level of abstraction. Many students come to not expect their work with fractions to make sense. Teachers have to provide a more active and direct role in providing relevant experiences with fractions.

- Young children's experiences in sharing equal amounts can provide a good introduction to fractions. Dividing a whole into equal shares opens the concept of equivalent fractions.

- Instruction in fractions that is rule-based and highly dependent on memorizing formulas may result in a rapid deterioration of proficiency as students forget the formulas.

---

[1] See National Research Council, *Adding It Up: Helping Children Learn Mathematics* (2001) for a more complete summary.

Models for Reteaching Fraction Concepts

Following are a series of models that teachers can use to reteach concepts at different stages of understanding fractions. These models expand on the recommended teaching sequence for fractions, which is as follows:

1. Define fractions.

2. Model fractions.

3. Compare fractions.

4. Multiply fractions.

5. Model the equivalence of fractions.

6. Simplify fractions.

7. Add and subtract fractions.

8. Divide fractions.

Most of these models are designed to use pattern blocks. Using pattern blocks not only helps students visualize and manipulate fractional relationships to understand fractions instead of simply trying to memorize formulas, but it also produces enormous benefits for a better understanding of algebraic ideas.

Model:
Defining Fractions

Recognizes that a fraction is a number that references a part or parts (numerator) of a whole (denominator).

Introduce the concept of a fraction as a ratio of equal parts to a whole.

Use pattern blocks for the following activities.

Suppose we define a fraction in the following way: A *fraction* is a number in the form of a ratio of equal parts to the whole. A *ratio* is a comparison of two things. The *numerator* is the number of parts of the whole. The *denominator* is the total number of parts that make up the whole. A fraction with the same numeral in both the numerator and denominator ($\frac{3}{3}$, for example) is equal to 1, since the number of parts in the fraction is equal to the total number of parts in the whole.

Now let's model this definition of a fraction using pattern blocks.

whole $\frac{1}{2}$ $\frac{1}{3}$ $\frac{1}{6}$

Y = 1 $R = \frac{1}{2}$ $B = \frac{1}{3}$ $G = \frac{1}{6}$

Next, let's examine the relationships between these pattern blocks:

$$2\,R = Y \qquad 3\,B = Y \qquad 6\,G = Y$$

$$R = \tfrac{1}{2}Y \qquad B = \tfrac{1}{3}Y \qquad G = \tfrac{1}{6}Y$$

To state it in words:

Red is one of the *two* equal blocks that it takes to make the yellow block.

Blue is one of the *three* equal blocks that it takes to make the yellow block.

Green is one of the *six* equal blocks that it takes to make the yellow block.

This models the definition of a fraction.

Have students practice modeling different fractions of the same whole.

Practice

Use pattern blocks to model the following. Be sure to define the whole (1).

1. $\frac{1}{2}$ 2. $\frac{1}{3}$ 3. $\frac{1}{4}$

4. $\frac{2}{3}$ 5. $\frac{4}{6}$ 6. $\frac{3}{4}$

7. $\frac{3}{2}$ 8. $\frac{4}{3}$ 9. $\frac{1}{6}$

10. $\frac{3}{6}$ 11. $\frac{5}{6}$ 12. $\frac{5}{12}$

Model:	Orders fractions with common and unlike denominators,
Comparing Fractions	demonstrating an understanding of how the numerator and denominator affect the value of fractions.

Use pattern blocks to compare fractional values and determine which fraction is greater or lesser.

Demonstrate and then have students use pattern blocks to compare the following fractions. Have them think about where the fractions would be placed on a number line. Some fractions have unlike denominators, which will encourage students to think.

Have students use the less than (<) or greater than (>) signs to show the relationships between the two fractions. Remember to define the whole (1) in the denominator.

Practice

< Less than   > Greater than

1. $\frac{1}{3}$ $\frac{2}{3}$ **2.** $\frac{3}{6}$ $\frac{5}{6}$ **3.** $\frac{1}{5}$ $\frac{3}{5}$

4. $\frac{1}{4}$ $\frac{3}{4}$ **5.** $\frac{1}{3}$ $\frac{1}{2}$ **6.** $\frac{2}{5}$ $\frac{1}{4}$

7. $\frac{3}{8}$ $\frac{3}{4}$ **8.** $\frac{5}{12}$ $\frac{1}{3}$ **9.** $\frac{2}{3}$ $\frac{1}{2}$

10. $\frac{1}{4}$ $\frac{2}{3}$ **11.** $\frac{1}{4}$ $\frac{1}{8}$

Review with students the relationships of the blocks to each other and to the whole, as defined.

Let's look at the relationships between blocks again. Demonstrate and then have students model each of the following "if/then" relationships with pattern blocks.

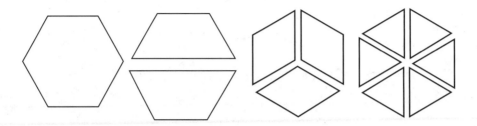

Assume Y = 1 for each of the following statements:

If R = $\frac{1}{2}$Y,   then $\frac{1}{2}$(1) = $\frac{1}{2}$.

If B = $\frac{1}{3}$Y,   then $\frac{1}{3}$(1) = $\frac{1}{3}$.

If G = $\frac{1}{6}$Y,   then $\frac{1}{6}$(1) = $\frac{1}{6}$.

If 2B = $\frac{2}{3}$Y,   then $\frac{2}{3}$(1) = $\frac{2}{3}$.

If 3B = $\frac{3}{3}$Y,   then $\frac{3}{3}$(1) = $\frac{3}{3}$ = 1.

If 2G = 1B,   then $\frac{2}{6}$Y = $\frac{1}{3}$Y   and $\frac{2}{6}$ = $\frac{1}{3}$.

If 3G = 1R,   then $\frac{3}{6}$Y = $\frac{1}{2}$Y   and $\frac{3}{6}$ = $\frac{1}{2}$.

If 4G = 2B,   then $\frac{4}{6}$Y = $\frac{2}{3}$Y   and $\frac{4}{6}$ = $\frac{2}{3}$.

If 6G = 3B = 2R, then   6 $(\frac{1}{6})$ = 3 $(\frac{1}{3})$ = 2$(\frac{1}{2})$ and $\frac{6}{6}$ = $\frac{3}{3}$ = $\frac{2}{2}$.

Use wholes of different shapes and sizes so that students recognize that any whole can be divided into fractions.

If this shape (2Y) = 1, then:

This shape = 4R and R = $\frac{1}{4}$.

This shape = 6B and B = $\frac{1}{6}$.

This shape = 12G and G = $\frac{1}{12}$.

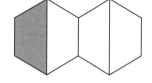

Demonstrate the following using pattern blocks:

1. The whole (1) =

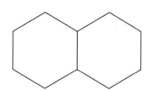

2. 2R = $\frac{1}{2}$

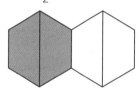

3. 3R = $\frac{3}{4}$

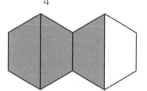

4. 5R = $\frac{5}{4}$ = 1 $\frac{1}{4}$

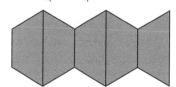

5. 3B = $\frac{1}{2}$

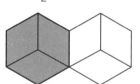

6. 4B = $\frac{4}{6}$ = $\frac{2}{3}$

7. 6B = 1

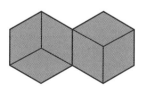

8. 6G = $\frac{1}{2}$

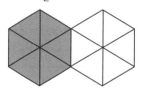

9. 5G = $\frac{5}{12}$

10. 8G = $\frac{8}{12}$ = $\frac{4}{6}$ = $\frac{2}{3}$

Assess students.

Ask students to write equivalent fractions for $\frac{3}{4}$ ($\frac{6}{8}$, $\frac{9}{12}$, $\frac{15}{20}$).

Model:
Multiplying Fractions by Whole Numbers

Understands that the fractional form of a whole number is $\frac{n}{1}$, and computes the product of two fractions by multiplying numerators and denominators.

Introduce multiplying fractions by whole numbers.

Let ⬡ = 1

A.

B.

C.

D.

E.

F.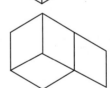

Which of the above figures model the following equations?

1. $2 \times \frac{1}{6} = \frac{2}{6}$ or $\frac{1}{3}$

2. $2 \times \frac{1}{2} = \frac{2}{2} = 1$

3. $2 \times \frac{1}{3} = \frac{2}{3}$

4. $3 \times \frac{1}{2} = \frac{3}{2}$ or $1\frac{1}{2}$

5. $2 \times \frac{2}{3} = \frac{4}{3}$ or $1\frac{1}{3}$

6. $4 \times \frac{1}{6} = \frac{4}{6}$ or $\frac{2}{3}$

Ask students:

Can the answer be found without using pattern blocks?

Do you see that $a \times \frac{b}{c} = \frac{a}{1} \times \frac{b}{c} = \frac{a \times b}{c}$?

Another way to state it is $3 \times \frac{1}{6} = \frac{3 \times 1}{1 \times 6} = \frac{3}{6} = \frac{1}{2}$.

That's the pattern!

| Model: **Multiplying Fractions by Fractions** | Computes the product of two fractions by multiplying numerators and denominators. |

Introduce multiplying fractions by fractions.

What is $\frac{1}{2}$ of $\frac{1}{3}$? (Note that $\frac{1}{2}$ of $\frac{1}{3}$ means $\frac{1}{2} \times \frac{1}{3}$.)

Let 1 =

$$\frac{1}{2} \times \frac{1}{3} = \frac{1}{6}$$

What is $\frac{1}{2}$ of $\frac{2}{3}$?

$\frac{2}{3}$ $\frac{1}{3}$

If 1 =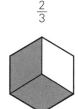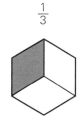

$$\frac{1}{2} \text{ of } \frac{2}{3} = \frac{1}{3}$$

What is $\frac{1}{2}$ of $\frac{2}{3}$?

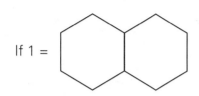

If 1 =

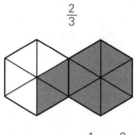

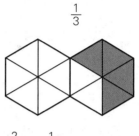

$\frac{1}{2}$ of $\frac{2}{3}$ = $\frac{4}{12}$ or $\frac{2}{6}$ or $\frac{1}{3}$

What is $\frac{1}{3}$ of $\frac{5}{6}$?

To summarize, let's look at a step-by-step approach to solving this problem.

Step 1:

If this is the whole (1),

then this is $\frac{5}{6}$.

But these 5 equal parts cannot be divided into 3 equal parts or $\frac{1}{3}$.

Step 2:

Suppose, instead, that this is the whole (1):

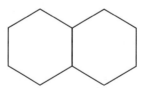

Then the shaded area shows $\frac{5}{6}$ of the whole.

But this $\frac{5}{6}$ cannot be divided into 3 equal parts either.

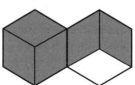

Step 3:

So let's suppose that this is the whole (1):

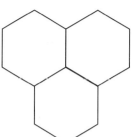

The shaded area of this figure is $\frac{5}{6}$, or $\frac{15}{18}$ of the whole.

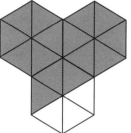

While the 5 equal sections shown in Step 2 cannot be divided into 3 equal parts, the 15 equal parts shown here can be. 15 equal parts divided by 3 is 5 equal parts.

That is, 5 of the 18. Therefore, we can say that $\frac{1}{3} \times \frac{5}{6} = \frac{15}{18}$.

Ask students: Do you notice a pattern?

Do you see that *to get the product of two fractions, you multiply numerator by numerator and denominator by denominator?*

$$\frac{a}{b} \times \frac{c}{d} = \frac{ac}{bd}$$

Assess students.

Ask students to compute $\frac{2}{3} \times \frac{3}{4}$ or, stated another way, what is $\frac{2}{3}$ of $\frac{3}{4}$? ($\frac{1}{2}$)

Model:
Equivalent Fractions

Recognizes that two fractions, such as $\frac{1}{3}$ and $\frac{2}{6}$, are equal when they represent the same portion of a whole.

Demonstrate equivalent fractions using pattern blocks.

1. Take a yellow pattern block (the hexagon).

2. Divide it into 2 equal parts. Use the red pattern blocks.
2 reds = 1 yellow

3. Divide the yellow block into 3 equal parts.
You'll use the blue blocks.
3 blue = 1 yellow

4. Then divide the yellow block into 6 equal parts.
You'll use the green triangles.
6 green = 1 yellow

Now let's recall the relationships between the blocks.

If 2 red = 1 yellow, 1 red is $\frac{1}{2}$Y.

If 3 blue = 1 yellow, 1 blue is $\frac{1}{3}$Y.

If 6 green = 1 yellow, 1 green is $\frac{1}{6}$Y.

But there are other relationships as well. For example:

3 green = 1 red

2 green = 1 blue

So if yellow represents 1, red = $\frac{1}{2}$, blue = $\frac{1}{3}$, green = $\frac{1}{6}$, 2 green = $\frac{2}{6}$, and 3 green = $\frac{3}{6}$,

then: $\frac{1}{2} = \frac{3}{6}$ $\frac{1}{3} = \frac{2}{6}$ $\frac{2}{2} = \frac{3}{3} = \frac{6}{6} = 1$

Have students practice modeling equivalent fractions.

Model the following equivalent fractions with pattern blocks.

Let 1 =

1.
$$\frac{1}{4} = \frac{3}{12}$$

2.
$$\frac{1}{3} = \frac{4}{12}$$

3.
$$\frac{1}{6} = \frac{2}{12}$$

4.
$$\frac{1}{2} = \frac{2}{4}$$

5.
$$\frac{3}{4} = \frac{9}{12}$$

(continued)

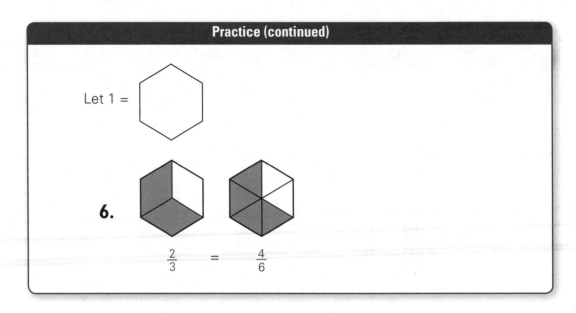

Let 1 =

6.

$$\frac{2}{3} = \frac{4}{6}$$

Develop an algorithm for finding equivalent fractions.

Say to students: Look at the equations above. Do you see the pattern?

$$\frac{a}{b} = \frac{ca}{cb}$$

That is, to change $\frac{1}{2}$ to fourths, multiply by $\frac{2}{2}$. In other words, $\frac{2}{2} \times \frac{1}{2} = \frac{2}{4}$.

To change $\frac{3}{6}$ to halves, divide by $\frac{3}{3}$: $\frac{3 \div 3}{6 \div 3} = \frac{1}{2}$.

Using the pattern illustrated above, how would you convert:

$\frac{1}{2}$ to sixths, or $\frac{1}{2} = \frac{?}{6}$

$\frac{1}{3}$ to sixths, or $\frac{1}{3} = \frac{?}{6}$

$\frac{2}{3}$ to sixths, or $\frac{2}{3} = \frac{?}{6}$

Answer: Multiply each fraction (numerator and denominator) by 1 in the form of $\frac{3}{3}$ or $\frac{2}{2}$.

In other words, use the Identity Property: Any fraction multiplied by 1 is itself ($1 \times a = a$).

$$\frac{3}{3} \times \frac{1}{2} = \frac{3}{6}$$

$$\frac{2}{2} \times \frac{1}{3} = \frac{2}{6}$$

$$\frac{2}{2} \times \frac{2}{3} = \frac{4}{6}$$

Now, demonstrate the above equations using pattern blocks.

Next, have students look at the relationship of individual pattern blocks to a whole defined as two hexagons.

Let 1 (whole) =

What fractional part of this whole would 1 red be? ($\frac{1}{4}$)

What fractional part would one blue be? ($\frac{1}{6}$)

What fractional part would one green be? ($\frac{1}{12}$)

Assess students.

Have students find equivalent fractions for $\frac{2}{3}$, $\frac{3}{4}$, $\frac{5}{6}$, $\frac{1}{3}$.

Mixed Number Equivalence

Understand that a mixed number such as $3\frac{2}{5}$ represents the sum of a whole number and a fraction less than one.

Introduce the concept of mixed fractions using models.

Demonstrate the following equations with pattern blocks, as shown:

If Y = 1, then …

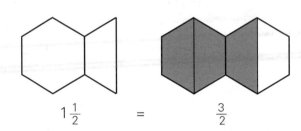

$$1\frac{1}{2} \qquad = \qquad \frac{3}{2}$$

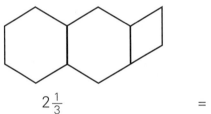 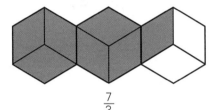

$$2\frac{1}{3} \qquad = \qquad \frac{7}{3}$$

Ask students: Do you see a pattern?

$2\frac{1}{3} = 2 + \frac{1}{3}$, or $\frac{3}{3} + \frac{3}{3} + \frac{1}{3} = \frac{7}{3}$, or $\frac{3 \times 2 + 1}{3}$.

Assess students.

Have students convert $2\frac{1}{4}$ to an improper fraction. $(\frac{9}{4})$

Simplifying Fractions

Simplifies or reduces fractions so that the numerator and denominator have no common factors other than 1: $\frac{6}{12} = \frac{3}{6} = \frac{1}{2}$.

Introduce the algorithm for simplifying fractions, and then model the concept with pattern blocks to develop understanding.

Let's examine the traditional method of simplifying fractions. Say we want to simplify the fraction $\frac{8}{12}$. The following is one method customarily used:

$$\frac{8}{12} = \frac{2 \times 4}{2 \times 6} = \frac{2 \times 2 \times 2}{2 \times 2 \times 3} = \frac{2}{2} \times \frac{2}{2} \times \frac{2}{3} = 1 \times 1 \times \frac{2}{3} = \frac{2}{3}.$$

Here's another method: $\frac{8}{12} = \frac{8 \div 4}{12 \div 4} = \frac{2}{3}$.

Either method gives us the answer. So what's the problem with it?

The problem is that it's "cookbook" math. Students often learn this system for simplifying fractions without *understanding* it. This book proposes a different method of instruction.

Let's begin with a concrete model of the problem using pattern blocks. We want to model $\frac{8}{12}$. How can we do that? Suppose we select 12 green pattern blocks. Why 12? Because the denominator tells us how many equal parts there are in the whole. So we have:

$$= \frac{8}{12}$$

$$= \frac{4}{6}$$

$$= \frac{2}{3}$$

What we have just modeled is $\frac{8}{12} = \frac{4}{6}$, if grouping by twos. If grouping by fours, then $\frac{8}{12} = \frac{2}{3}$.

Let's look at another example: Change $\frac{6}{15}$ to other forms.

Select 15 pattern blocks of the same color. Suppose we chose blue.

$$\frac{6b}{15b} \qquad \frac{b\,b\,b\,b\,b\,b}{b\,b\,b\,b\,b\,b\,b\,b\,b\,b\,b\,b\,b\,b\,b}$$

Now, let's arrange these blocks in the smallest number of sets possible but with the same number on the top as on the bottom:

$$\frac{bbb\ bbb}{bbb\ bbb\ bbb\ bbb\ bbb} = \frac{2}{5}$$

Answer: $\frac{6}{15} = \frac{2}{5}$ or $\frac{6 \div 3}{15 \div 3} = \frac{2}{5}$.

Remember that math is the science of patterns and relationships, so let's look for the pattern and develop (discover) an algorithm (rule). After several models are demonstrated, students will usually discover the traditional algorithm for simplifying fractions.

To raise a fraction to higher terms, simply reverse the process. For example, suppose we want to demonstrate that $\frac{1}{2} = \frac{3}{6}$.

We can show the fraction using pattern blocks. Suppose we use yellow.

Remember that a fraction is a number in the form of a ratio (comparison) of equal parts to the whole.
Let's assume the whole is 2Y in this case.

So 1 of the 2 is Y.

But Y = 2R = 3B = 6G.

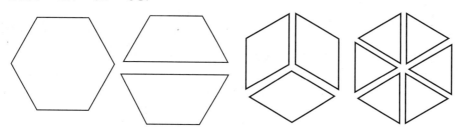

$\frac{Y}{YY} = \frac{2R}{4R} = \frac{3B}{6B} = \frac{6G}{12G}$. So $\frac{1}{2} = \frac{2}{4} = \frac{3}{6} = \frac{6}{12}$. So $\frac{1}{2} = \frac{3}{6}$, which is what we want to demonstrate. If we want to change $\frac{1}{2}$ to sixths, we can multiply both the numerator and the denominator by 3:

$$\frac{1}{2} \times \frac{3}{3} = \frac{3}{6}$$

Assess students.

Have students simplify $\frac{15}{18}$ using pattern blocks and the traditional algorithm.

| Model:
Adding and Subtracting Fractions with Common Denominators | Understands that fractions with the same denominator can be added or subtracted by adding or subtracting the numerators. |

Model adding and subtracting fractions with like denominators using pattern blocks.

Model the following equations with pattern blocks, and have students look for a pattern.

1. $\frac{1}{2} + \frac{1}{2} = \frac{2}{2} = 1$

2. $\frac{1}{3} + \frac{1}{3} = \frac{2}{3}$

3. $\frac{1}{6} + \frac{1}{6} = \frac{2}{6}$

4. $\frac{1}{6} + \frac{2}{6} = \frac{3}{6} = \frac{1}{2}$

5. $\frac{2}{6} + \frac{3}{6} = \frac{5}{6}$

6. $\frac{5}{12} + \frac{1}{12} = \frac{6}{12} = \frac{1}{2}$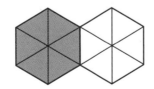

Ask students: What's the pattern or rule? Did you notice that the denominator always remains the same because the numerator indicates the addition of more parts of the whole?

Model the following and have students look for the pattern for subtracting fractions with common denominators.

1. $\frac{2}{3} - \frac{1}{3} = \frac{1}{3}$

2. $\frac{3}{6} - \frac{1}{6} = \frac{2}{6} = \frac{1}{3}$

3. $\frac{4}{6} - \frac{1}{6} = \frac{3}{6} = \frac{1}{2}$

Summarize the rule for adding and subtracting fractions with common denominators.

Have students summarize how to add or subtract fractions with the same denominator.

(Answer: When you add or subtract fractions with the same denominator, you add or subtract the numerator and keep the same denominator.)

Have students practice modeling the following equations with pattern blocks.

Model the following equations with pattern blocks. Assume Y = 1.

a. $\frac{1}{3} + \frac{1}{3} = \frac{2}{3}$

b. $\frac{1}{2} + \frac{1}{3} = \frac{5}{6}$

c. $\frac{1}{2} + \frac{1}{2} = \frac{2}{2} = 1$

d. $\frac{1}{2} + \frac{1}{6} = \frac{4}{6} = \frac{2}{3}$

e. $\frac{1}{6} + \frac{1}{6} = \frac{2}{6} = \frac{1}{3}$

f. $\frac{1}{3} + \frac{1}{6} = \frac{3}{6} = \frac{1}{2}$

Assess students.

Ask students to mentally subtract $\frac{2}{5}$ from $\frac{4}{5}$.

Adding Fractions with Unlike Denominators | Adds fractions with unlike denominators by finding a common denominator.

Model adding fractions with unlike denominators, recognizing the need to find equivalent fractions with common denominators.

Model the following and have students find the pattern.

1. $\frac{1}{2} + \frac{1}{3} = \frac{5}{6}$

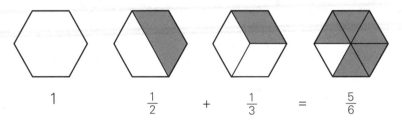

1 $\qquad$ $\frac{1}{2}$ $\quad +\quad$ $\frac{1}{3}$ $\quad =\quad$ $\frac{5}{6}$

2. $\frac{1}{6} + \frac{1}{2} = \frac{4}{6} = \frac{2}{3}$

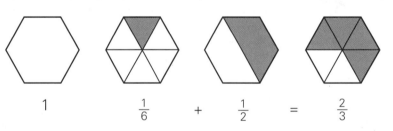

1 $\qquad$ $\frac{1}{6}$ $\quad +\quad$ $\frac{1}{2}$ $\quad =\quad$ $\frac{2}{3}$

3. $\frac{2}{3} + \frac{1}{2} = \frac{7}{6} = 1\frac{1}{6}$

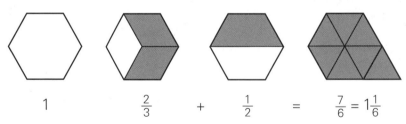

1 $\qquad$ $\frac{2}{3}$ $\quad +\quad$ $\frac{1}{2}$ $\quad =\quad$ $\frac{7}{6} = 1\frac{1}{6}$

4. $\frac{2}{3} + \frac{1}{6} = \frac{5}{6}$

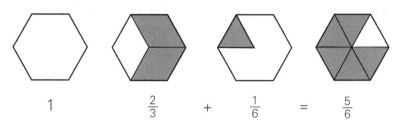

$\qquad$ 1 $\qquad\qquad$ $\frac{2}{3}$ $\qquad$ + $\qquad$ $\frac{1}{6}$ $\qquad$ = $\qquad$ $\frac{5}{6}$

5. $\frac{3}{4} + \frac{1}{2} = \frac{5}{4} = 1\frac{1}{4}$

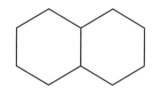

1

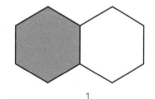 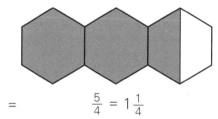

$\qquad$ $\frac{3}{4}$ $\qquad$ + $\qquad$ $\frac{1}{2}$ $\qquad$ = $\qquad$ $\frac{5}{4} = 1\frac{1}{4}$

6. $\frac{1}{8} + \frac{3}{4} = \frac{7}{6}$

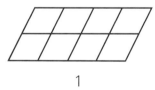

1

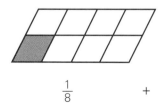 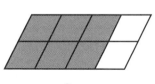 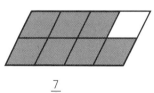

$\qquad$ $\frac{1}{8}$ $\qquad$ + $\qquad$ $\frac{3}{4}$ $\qquad$ = $\qquad$ $\frac{7}{8}$

Have students practice adding fractions with unlike denominators.

Model these equations with pattern blocks.

1. $\frac{1}{6} + \frac{1}{2} = \frac{1}{6} + \frac{3}{6} = 1 + \frac{3}{6} = \frac{4}{6} = \frac{2}{3}$

2. $\frac{2}{5} + \frac{1}{2} = \frac{4}{10} + \frac{5}{10} = \frac{4+5}{10} = \frac{9}{10}$

3. $\frac{2}{3} + \frac{1}{6} = \frac{4}{6} + \frac{1}{6} = \frac{4+1}{6} = \frac{5}{6}$

4. $\frac{2}{7} + \frac{1}{2} = \frac{4}{14} + \frac{7}{14} = \frac{4+7}{14} = \frac{11}{14}$

5. $\frac{3}{4} + \frac{1}{2} = \frac{3}{4} + \frac{2}{4} = \frac{3+2}{4} = \frac{5}{4} = 1\frac{1}{4}$

6. $\frac{1}{8} + \frac{3}{4} = \frac{1}{8} + \frac{6}{8} = \frac{1+6}{8} = \frac{7}{8}$

7. $\frac{1}{9} + \frac{2}{3} = \frac{1}{9} + \frac{6}{9} = \frac{1+6}{9} = \frac{7}{8}$

8. $\frac{2}{5} + \frac{3}{8} = \frac{16}{40} + \frac{11}{14} = 16 + \frac{16+15}{40} = \frac{31}{40}$

Ask students: Do you see the pattern? Can you state it?

(Answer: $\frac{a}{b} + \frac{c}{d} + \frac{ad+cb}{bd}$ or $\frac{d}{d} \times \frac{a}{b} + \frac{b}{b} \times \frac{c}{d}$)

Assess students.

Have students state the pattern for adding fractions with unlike denominators in their own words.

Dividing Fractions | Divides fractions using multiplication by the inverse.

Introduce the concept of dividing fractions with pattern blocks.

Recall that a fraction is a number in the form of a ratio of *equal parts* to the whole. Therefore, the division of fractions is the ratio of the two fractions. *Ratio* is the comparison of equal parts.

Model the following:

1. $\dfrac{\frac{1}{2}}{\frac{1}{3}} = ?$

Yellow   Red   Blue

1 $\dfrac{1}{2}$ $\div$ $\dfrac{1}{3}$

Here's the answer:

What's the value?

$$\dfrac{\;\;}{\;\;} \qquad \dfrac{\frac{1}{2}}{\frac{1}{3}} = \dfrac{\;\;}{\;\;} = \dfrac{3}{2}$$

$$\dfrac{\frac{1}{2}}{\frac{1}{3}} = \dfrac{3}{2}$$

2. $\dfrac{\frac{1}{3}}{\frac{1}{2}} = ?$

To solve, compare equal parts:

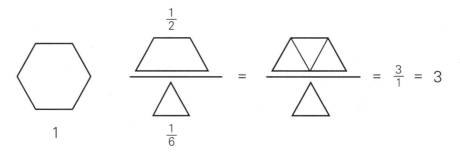

the ratio of
equal parts

3. $\dfrac{\frac{1}{2}}{\frac{1}{6}} = ?$

$$\frac{\frac{1}{2}}{\frac{1}{6}} \quad = \quad \quad = \quad \frac{3}{1} \quad = \quad 3$$

4. $\dfrac{\frac{2}{3}}{\frac{1}{4}} = ?$

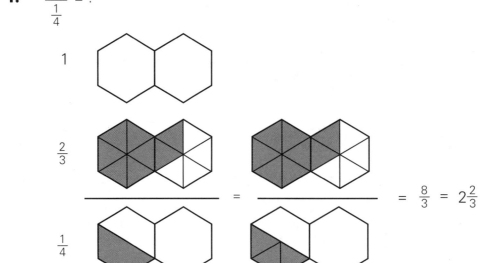

Introduce algorithms for dividing fractions.

There are two ways to find the answer to the division of fractions without using pattern blocks. Both of them involve using algorithms. For example:

a. Because $1 \times A = A$,

$$\frac{\frac{1}{3}}{\frac{1}{2}} = \frac{6}{6} \times \frac{\frac{1}{3}}{\frac{1}{2}} = \frac{2}{3}$$

Multiply by the least common denominator in the form of $\frac{a}{a}$.

b. Because $\frac{A}{A} = 1$,

$$\frac{\frac{1}{3}}{\frac{1}{2}} = \frac{1}{3} \times \frac{2}{1} = \frac{2}{3}$$

Invert and multiply numerators and denominators. Or stated in its expanded form:

$$\frac{\frac{2}{1} \times \frac{1}{3}}{\frac{2}{1} \times \frac{1}{2}} = \frac{\frac{2}{1} \times \frac{1}{3}}{1} = \frac{2}{1} \times \frac{1}{3} = \frac{2}{3}$$

Chapter Wrap-Up

Practice Ideas

Fractions on a Number Line Reinforce the concept of fractions as numbers and the value of fractions by having students place fractions on a number line, stacking equivalent fractions.

Equivalent Fractions Reinforce finding equivalent fractions and simplifying fractions by giving students a fraction, such as $\frac{2}{3}$ or $\frac{4}{16}$, and seeing how many equivalent fractions they can record in one minute.

Fraction Addition and Subtraction Practice Provide periodic opportunities to maintain skills by asking students to solve addition and subtraction problems that require:

- addition and subtraction of fractions with common denominators

- addition and subtraction of fractions with unlike denominators

- addition and subtraction of mixed and improper fractions that need to be regrouped

Multiplication and Division of Fractions Provide periodic practice opportunities to maintain skills and understanding of multiplying and dividing fractions by asking students to solve problems that require:

- multiplication and division of fractions by whole numbers

- division of fractions by fractions

- multiplication of fractions by fractions

- multiplication and division of mixed fractions

Further Reteaching

The teaching models in this chapter demonstrate how teachers can use manipulatives and visual models to reteach fractions concepts. Because student understanding of fractions suffers from a lack of real-world experience with fractions, using manipulatives to reteach concepts is a very effective method for struggling students to visualize fractions and operations with fractions. Having students demonstrate their understandings with manipulatives is an excellent way to confirm that they have internalized the concepts and are not just trying to memorize formulas and guess at answers without understanding.

Skill Maintenance and Assessment

Use the assessments on the following pages to assess and refresh student understanding of operations with fractions.

Adding Fractions

1a. $\frac{1}{7} + \frac{1}{12} =$ ☐

1b. $\frac{4}{5} + \frac{2}{3} =$ ☐

1c. $\frac{1}{6} + \frac{2}{8} =$ ☐

2a. $\frac{3}{6} + \frac{1}{11} =$ ☐

2b. $\frac{6}{8} + \frac{3}{4} =$ ☐

2c. $\frac{7}{11} + \frac{6}{7} =$ ☐

3a. $\frac{8}{9} + \frac{1}{9} =$ ☐

3b. $\frac{3}{9} + \frac{7}{8} =$ ☐

3c. $\frac{3}{9} + \frac{3}{6} =$ ☐

4a. $\frac{2}{7} + \frac{9}{11} =$ ☐

4b. $\frac{1}{7} + \frac{7}{8} =$ ☐

4c. $\frac{1}{3} + \frac{5}{10} =$ ☐

Subtracting Fractions

1a. $\frac{7}{11} - \frac{1}{8} =$ ☐

1b. $\frac{8}{9} - \frac{5}{6} =$ ☐

1c. $\frac{5}{6} - \frac{2}{12} =$ ☐

2a. $\frac{10}{12} - \frac{8}{11} =$ ☐

2b. $\frac{9}{10} - \frac{6}{12} =$ ☐

2c. $\frac{6}{7} - \frac{5}{6} =$ ☐

3a. $\frac{2}{12} - \frac{1}{7} =$ ☐

3b. $\frac{7}{12} - \frac{2}{4} =$ ☐

3c. $\frac{7}{8} - \frac{4}{5} =$ ☐

4a. $\frac{10}{11} - \frac{6}{11} =$ ☐

4b. $\frac{3}{10} - \frac{1}{4} =$ ☐

4c. $\frac{2}{5} - \frac{2}{6} =$ ☐

Multiplying Fractions

1a. $\frac{1}{2} \times \frac{3}{7} =$ ☐

1b. $\frac{1}{2} \times \frac{2}{3} =$ ☐

1c. $\frac{5}{6} \times \frac{1}{7} =$ ☐

2a. $\frac{2}{3} \times \frac{2}{3} =$ ☐

2b. $\frac{7}{9} \times \frac{5}{10} =$ ☐

2c. $\frac{3}{5} \times \frac{1}{6} =$ ☐

3a. $\frac{8}{11} \times \frac{1}{6} =$ ☐

3b. $\frac{3}{4} \times \frac{1}{8} =$ ☐

3c. $\frac{1}{4} \times \frac{1}{7} =$ ☐

4a. $\frac{1}{2} \times \frac{1}{3} =$ ☐

4b. $\frac{1}{6} \times \frac{3}{4} =$ ☐

4c. $\frac{8}{9} \times \frac{10}{11} =$ ☐

Dividing Fractions

1a. $\frac{1}{6} \div \frac{2}{5} =$ ☐

1b. $\frac{7}{11} \div \frac{1}{4} =$ ☐

1c. $\frac{1}{3} \div \frac{2}{3} =$ ☐

2a. $\frac{1}{6} \div \frac{9}{11} =$ ☐

2b. $\frac{3}{4} \div \frac{1}{4} =$ ☐

2c. $\frac{3}{7} \div \frac{1}{12} =$ ☐

3a. $\frac{1}{4} \div \frac{5}{11} =$ ☐

3b. $\frac{1}{5} \div \frac{3}{5} =$ ☐

3b. $\frac{7}{11} \div \frac{5}{6} =$ ☐

4a. $\frac{2}{3} \div \frac{4}{5} =$ ☐

4b. $\frac{1}{2} \div \frac{2}{3} =$ ☐

4c. $\frac{1}{6} \div \frac{1}{3} =$ ☐

Chapter 4

Decimals

Chapter Introduction

What Are Decimals?

Before introducing one or more of the reteaching models proposed in this chapter, you will probably want to review with your students the following concepts, vocabulary, and properties of operations with decimals.

A decimal (0.5) is really a fraction. Its denominator is a power of 10. For example, 0.5 is $\frac{5}{10}$, which can be simplified to $\frac{1}{2}$. The number 2.75 is the mixed number 2 and $\frac{75}{100}$, which can be simplified to $2\frac{3}{4}$.

Decimals are based on the base-10 system. Most people are quite familiar with using decimals when dealing with money. In the amount $10.25, the 0.25 refers to 25 of 100 cents.

Decimals can be converted to fractions and fractions can be converted to decimals. Some decimal fractions are *irrational*, which means that they can be calculated without end and with no pattern. The number pi (π) is an irrational number. Its value cannot be expressed exactly as a fraction and consequently its decimal representation never ends or repeats. It has been calculated to over one trillion digits with no recognizable pattern!

On a number line, decimals (like fractions) are the numbers that come between the *integers* or *whole numbers*. For example, 0.35 is between 0 and 1 on a number line.

Vocabulary of Decimals

A *decimal fraction* is a fraction whose denominator is a power of 10 (for example, $\frac{1}{10}$, $\frac{33}{100}$, $\frac{44}{1000}$).

A *decimal point* is the period that separates the whole number or integer from the fractional part that follows. When we read a decimal, the decimal point is read as "and." (For example, 32.5 is read as "thirty-two *and* 5 tenths.")

Properties of Operations with Decimals

Operations with decimals use the same properties as do whole numbers (the associative, commutative, and distributive properties).

Operations with decimals have certain properties	
Commutative Property	The order in which two numbers are added or multiplied does not matter: $5.4 \times 4 = 4 \times 5.4$.
Associative Property	When three or more numbers are added or multiplied, the product is the same regardless of the grouping of the factors: $(3.25 \times 3) \times 4 = 3.25 \times (3 \times 4)$.
Distributive Property	The factors in multiplication and division can be decomposed and operated on individually and then added together: $(3.4 \times 4) = (3 \times 4) + (0.4 \times 4)$.
Identity Property	Any number multiplied by 1 is itself: $0.33 \times 1 = 0.33$.
Zero Property	Anything multiplied by 0 is 0: $0.5 \times 0 = 0$.

Expectations for Middle School

Decimals are introduced in discussions of money as early as kindergarten. Decimal fractions and fractions are reinforced in the early grades in lessons on time (fractions of an hour), money (fractions of a dollar), and measurement ($\frac{1}{2}$ foot, $\frac{1}{2}$ mile). Adding and subtracting decimals is introduced as early as grade 4 and causes significant problems for some students, who attempt to apply their whole number understanding to the addition and subtraction and, later, multiplication and division of decimals.

By sixth grade, students should have a clear understanding of what decimals are, how they are different from whole numbers, and how they relate to fractions, percents, ratios, and proportion. They should be fluent in adding and subtracting decimals. In sixth grade they are expected to develop fluency in multiplying and dividing decimals.

Common Problems Students Have with Decimals

Students learn about decimals in stages and can get stuck in any stage. The basic stages of learning about decimals and the difficulties associated with those stages are listed following.

Learning Stage	Difficulty
Multiplying Decimals by 10 Is aware of the effect of multiplying by 10 on the location of the decimal point in a number.	*Unless students recognize that the decimal point location changes when multiplying a decimal by 10 or 100, they do not understand the significance of the decimal point in identifying the value of the decimal.*
Adding and Subtracting Decimals to 2 Places Applies the addition and subtraction algorithms (methods) to money.	*Until students can add and subtract amounts of money with decimals, they will not be able to develop the concept of decimal fractions.*
Ordering Decimals Orders whole numbers (integers) and positive fractions and decimals on a number line.	*If students cannot order decimals and fractions between integers in order of value, then they do not understand the value a decimal represents.*
Decimals to the Hundredths and Thousandths Place Reads and writes decimals to the thousandths place.	*If students do not extend their knowledge of decimals beyond money concepts with two-digit decimals, they may not have grasped the concept of decimals apart from money.*
Repeating and Terminating Decimals Identifies repeating and terminating decimals such as $\frac{1}{9}$ (0.11111…) and $\frac{1}{8}$ (0.125).	*Unless students recognize that some decimals repeat, they do not understand how decimals can represent values differently from fractions.*
Multiplying and Dividing Decimals to 2 Places Multiplies and divides decimals to two places and places the decimal point correctly.	*Unless students can place the decimal in the right place after multiplying or dividing, they do not understand the value of decimals and may be calculating mindlessly.*

Using Decimals Reads, writes, and compares numbers expressed as decimals and understands that a digit in one place represents ten times what it represents in the place to its right.	*Until students recognize the value of each digit in a decimal, they will not be able use decimals effectively.*
Decimal Equivalents Knows the decimal equivalents for unit fractions $\frac{1}{2}$, $\frac{1}{3}$, $\frac{1}{4}$, $\frac{1}{5}$, $\frac{1}{8}$, and $\frac{1}{9}$, and finds equivalent representations of fractions as decimals, ratios, and percentages.	*Until students can find the decimal equivalents for fractions, ratios, and percentages, they will not be able to gain fluency in working with fractions and decimals.*
Computing Decimals Computes sums, differences, products, and quotients of finite decimals using strategies based on place value, the properties of operations, and their inverse relationships.	*Until students can apply the properties of addition, subtraction, multiplication, and division to working with decimals, they will not gain fluency in using decimals to solve problems.*

What Research Says

Research has identified the following issues relating to the difficulty of understanding decimals.[1]

- Fractions are represented in multiple ways (as fractions $\frac{1}{2}$, decimals 0.5, and percents 50%), and they are used in many ways (as parts of wholes, ratios, and quotients). This causes considerable confusion. Each form has many common uses in daily life that do not seem connected. Most students learn rules for translating between forms, but they often understand very little about what quantities the symbols represent and make frequent and nonsensical errors.

- Young children's experiences with money can provide a good introduction to decimals, but many students may have difficulty recognizing amounts of money as decimals.

- Instruction in calculating decimals tends to be rule-based and highly dependent on memorizing formulas and procedures (move the decimal point two places), resulting in a rapid deterioration of proficiency as students forget the formulas and procedures.

---

[1] See National Research Council, *Adding It Up: Helping Children Learn Mathematics* (2001), for a more complete summary.

Models for Reteaching Decimal Concepts

Following are four models that teachers can use to reteach concepts at different stages of understanding decimals. The first model focuses on decimal fractions and place value, the second on finding decimal fraction equivalents, the third on placing fractions and decimals on a number line, and the fourth on operations with decimals.

Model:
Place Value

Read and write decimals to the thousandths place.

Introduce the place value of decimal fractions.

The term *decimal* should really be *decimal fraction*. All decimals are fractions with denominators of 10 or some multiple of 10.

This is a model of decimal place value for the number 23.4

10s	1s	.	Tenths
2	3	.	4

This is a model of decimal place value for the number 1,234.567

1,000s	100s	10s	1s	.	Tenths	Hundredths	Thousandths
1	2	3	4	.	5	6	7

Have students practice decimal fraction place value.

Practice							
1,000s	100s	10s	1s	.	Tenths	Hundredths	Thousandths

Write these decimal fractions in the chart.

    1. 13.2   **2.** 13.34   **3.** 5,289.2

    4. 9,843.231   **5.** 0.001

Read decimal fractions.

When you read decimal fractions, you reinforce that they are fractions.

	Read as...
$3\frac{1}{10} = 3.1$, or $3 + \frac{1}{10}$	"three and one-tenth."
$40\frac{2}{100} = 40.02 = 40 + \frac{0}{10} + \frac{2}{100}$	"forty and two one-hundredths."
$16\frac{234}{1,000} = 16.234 =$ $16 + \frac{2}{10} + \frac{3}{100} + \frac{4}{1,000}$ or $\frac{234}{1,000}$	"sixteen and two-hundred-thirty-four-thousandths."
$4\frac{5}{10} = 4.5 = 4 + \frac{5}{10}$	"four and five-tenths."
$101\frac{11}{100} = 101.11 =$ $101 + \frac{1}{10} + \frac{1}{100}$ or $\frac{11}{100}$	"one hundred one and eleven-hundredths."

Assess reading decimal fractions.

Have students write the following decimal fractions in decimal form.

1. one hundred twenty-four and five hundred thirty-two thousandths

2. two hundred and two-thousandths

3. one thousand four hundred fifty-four and five-tenths

4. two hundred twenty-two and twenty-two hundredths

5. one thousand and one-thousandth

Model:
Decimal Equivalents

Know the decimal equivalents for unit fractions $\frac{1}{2}$, $\frac{1}{3}$, $\frac{1}{4}$, $\frac{1}{5}$, $\frac{1}{8}$, and $\frac{1}{9}$ and find equivalent representations of fractions as decimals, ratios, and percentages.

Introduce finding decimal fraction equivalents.

Fractions are really division problems. The fraction $\frac{1}{2}$ is the same as $1 \div 2$. When we divide 1 by 2, we get 0.5, which is the decimal equivalent of $\frac{1}{2}$.

What is the decimal equivalent of $\frac{1}{4}$? $(1 \div 4)$ or 0.25

What is the decimal equivalent of $\frac{5}{8}$? $(5 \div 8)$ or 0.625

Can you find your answers on the following fraction/decimal equivalence chart?

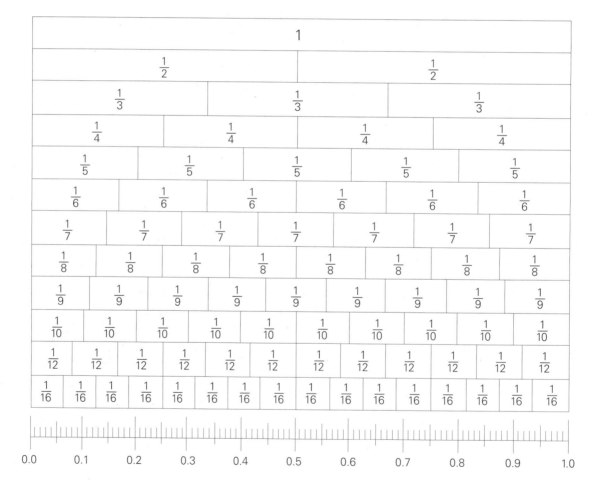

Have students practice finding equivalent decimal fractions.

<table>
<tr><td colspan="2">Practice</td></tr>
</table>

Practice changing the following from fractions to decimal fractions.

1. $\frac{1}{2} = \frac{5}{10} =$ **2.** $3\frac{6}{10} =$

3. $5\frac{40}{100} =$ **4.** $10\frac{5}{100} =$

5. $\frac{8}{10} =$ **6.** $\frac{12}{100} =$

7. $\frac{30}{1,000} =$ **8.** $\frac{66}{10,000} =$

9. $\frac{1}{100} =$ **10.** $4\frac{3}{10} =$

Model:
Number Line

Order whole numbers (integers) and positive fractions and decimals on a number line.

Demonstrate fractions and decimal fractions on a number line.

- Draw a number line from –10 to 10 on the board, or project the number line below on an interactive whiteboard. Plot 0 in the middle.

- Plot the following numbers on the number line: 5, 1, 2, –5.

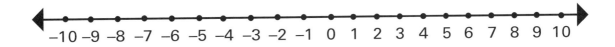

- Now plot $\frac{1}{2}$, $1\frac{1}{2}$, and $5\frac{1}{2}$ above the number line.
- Plot the decimal equivalents below the number line (0.5, 1.5, 5.5).
- Next, plot $\frac{1}{4}$ and $\frac{1}{8}$ above the number line and the decimal equivalents below (0.25, 0.125).
- Ask students: The number $\frac{1}{8}$ is small, but is it less than 0?

Have students practice plotting decimals and decimal fractions on a number line.

Plot these numbers as fractions and decimals on the number line.

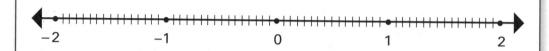

I. $\frac{1}{2}$ **2.** 1.25   **3.** 0.75

4. 1.125   **5.** $\frac{1}{16}$ =

Model:
Operations with Decimals

Compute sums, differences, products, and quotients of finite decimals using strategies based on place value, the properties of operations, and their inverse relationships.

Introduce adding and subtracting decimals.

One way to add or subtract decimals is to think of the decimals as decimal fractions:

$\frac{1}{10} + \frac{1}{100} =$ First, we find the common denominator:

$(\frac{10}{10} \times \frac{1}{10}) + \frac{1}{100} = \frac{10}{100} + \frac{1}{100} = \frac{11}{100}$

Similarly:
$$
\begin{array}{r}
0.1 \\
0.01 \\
\hline
\end{array} =
\begin{array}{r}
0.10 \\
+\ 0.01 \\
\hline
0.11
\end{array}
$$

Adding and subtracting decimals is a lot like adding and subtracting whole numbers. You have to remember to line up the numbers at the decimal point so they have the same place value.

You can also use Base-10 blocks to model adding and subtracting decimals. To do this, you will have to redefine the value of the blocks to match the decimal numbers. For example, if we let the 100 flat have a value of 1, then the ten-rod is $\frac{1}{10}$, and the unit block is $\frac{1}{100}$.

Have students practice adding and subtracting decimals.

<table>
<tr><td colspan="2" align="center">**Practice**</td></tr>
<tr><td colspan="2">Solve these problems. Use Base 10 blocks if necessary.</td></tr>
</table>

1.   3.16
     + 2.50

2.   4.02
     + 3.15

3.   6.39
     + 4.12

4.   10.002
     + 3.104

5.   4.63
     − 2.12

6.   9.35
     − 6.24

7.   10.01
     − 5.20

8.   8.31
     − 3.79

Introduce multiplying decimals.

Demonstrate that multiplying decimals is like multiplying whole numbers.

1.   Line up the numbers on the right. (Unlike addition and subtraction, you don't need to line up the decimal points.)

$$
\begin{array}{r}
3.77 \\
\times\ 2.8 \\
\hline
3.016 \\
+\ 754 \\
\hline
10.556
\end{array}
$$

  3.77   (2 decimal places)
× 2.8   (1 decimal place)

10.556   (3 decimal places)

2.   Multiply each digit in the top number by each digit in the bottom number, just as with whole numbers. Then add the products.

3.   Add together the decimal places in the two numbers you multiplied. Then place the decimal point in the answer by counting that number of places from the right.

4. Estimate your answer to make sure it makes sense:

Round 3.77 to 4 and 2.8 to 3.

Multiply 4×3. The answer 12 is close to 10.556. (If you had placed the decimal point in the wrong place, your answer would have been 1.0556 or 105.56, and your estimate would not have been close.)

You can also compare multiplying decimals to multiplying fractions.

What is $\frac{1}{10} \times \frac{1}{10}$? $\frac{1}{10} \times \frac{1}{10} = \frac{1}{100}$

What is 0.1×0.1? $0.1 \times 0.1 = 0.01$

What is $\frac{1}{10} \times \frac{2}{100}$? $\frac{1}{10} \times \frac{2}{100} = \frac{2}{1,000}$

What is 0.1×0.02? $0.1 \times 0.02 = 0.002$

The rule is to count decimal places and multiply as usual. (1 decimal place + 2 decimal places = 3 decimal places.)

Have students practice multiplying decimals.

Solve

Solve these problems.

1. $\begin{array}{r} 3.6 \\ \times\, 0.2 \\ \hline \end{array}$
    2. $\begin{array}{r} 5.01 \\ \times\, 0.31 \\ \hline \end{array}$

3. $\begin{array}{r} 4.5 \\ \times\, 3.6 \\ \hline \end{array}$
    4. $\begin{array}{r} 30.5 \\ \times\;\; 2.8 \\ \hline \end{array}$

5. $\begin{array}{r} 1.35 \\ \times\, 0.05 \\ \hline \end{array}$
    6. $\begin{array}{r} 100.6 \\ \times\, 0.004 \\ \hline \end{array}$

Introduce dividing decimals.

Let's look at dividing fractions first. Remember that you can simplify fractions.

$\frac{1}{10} \div \frac{1}{100}$ is the same as $\frac{1}{10} \times \frac{100}{1}$, or $\frac{100}{10}$, or $\frac{10}{1}$, or 10.

$$\frac{\frac{1}{10}}{\frac{1}{100}} = \frac{100}{100} \times \frac{\frac{1}{10}}{\frac{1}{100}} - \frac{100}{10} = 10$$

Another way to divide fractions is to multiply by the inverse ($\frac{1}{10} \times \frac{100}{1} = \frac{100}{10} = 10$).

$$0.01 \overline{).1}$$

Let's do the same thing as we did above by multiplying both the divisor and the dividend by 100 to make both of them whole numbers.

$$1 \overline{)10} \overset{10}{}$$

We can then proceed to divide the resulting whole numbers.

Let's look at another example:

$$3.8 \overline{)1{,}614.62}$$

$$38 \overline{)16{,}146.2}$$

When the divisor is not a whole number, move the decimal point to the right to make it a whole number and move the decimal point in the dividend the same number of places to the right. That's how you pay attention to place value.

What if there isn't a decimal in the dividend? Then add zero. That maintains the same ratio between the divisor and the dividend.

$$3.8 \overline{)161}$$

$$38 \overline{)1{,}610}$$

$$3.88 \overline{)161}$$

$$388 \overline{)16{,}100}$$

Have students practice dividing decimals.

Practice
Solve these problems.

1. $0.3\overline{)6.25}$ **2.** $0.38\overline{)69.34}$

3. $4.2\overline{)9.36}$ **4.** $0.02\overline{)4.39}$

5. $0.6\overline{)0.00382}$ **6.** $0.21\overline{)6.34}$

Assess computing decimals.

Have students solve these problems.

1. $0.8 + 0.42 =$

2. $1.08 + 2.3 =$

3. $6.40 - 0.58 =$

4. $1.08 - 0.3 =$

5. $0.8 \times 0.42 =$

6. $1.08 \times 2.3 =$

7. $0.58\overline{)6.40} =$

8. $0.3\overline{)1.08} =$

Practice Ideas

Decimals on a number line Reinforce the concept of decimals as numbers and the value of decimals by having students place decimals on a number line along with equivalent fractions.

Decimal addition and subtraction practice Provide periodic opportunities to maintain skills by asking students to solve addition and subtraction problems that require:

- Addition and subtraction of decimals with common denominators (3.5 + 2.3)

- Addition and subtraction of decimals with unlike denominators (3.05 – 2.3)

Decimal multiplication and division practice Provide periodic practice opportunities to maintain skills and understanding of multiplying and dividing decimals by asking students to solve problems that require:

- Multiplication and division of decimals by whole numbers

- Multiplication and division of decimals by decimals

Further Reteaching

The teaching models in this chapter demonstrate how teachers can use examples and visual models to reteach decimal concepts. Because student understanding of decimals suffers from a lack of real-world experience apart from money, using these models to reteach concepts can be a very effective way for struggling students to visualize decimals and operations with decimals. Having students demonstrate their understanding with Base-10 blocks is an excellent way to confirm that they have internalized the concepts and are not just trying to memorize formulas and guess at answers without understanding.

Skill Maintenance and Assessment

Use the exercises on the following page to assess and refresh student understanding of operations with decimals.

Operations with Decimals

1.
```
   0.062
−  0.710
```

2.
```
   0.062
+  0.710
```

3.
```
   0.153
−  2.820
```

4.
```
   0.153
+  2.820
```

5. $1.900\overline{)0.209}$

6. $1.8\overline{)16.2}$

7. $1.70\overline{)2.38}$

8.
```
    19.390
×  282.500
```

9. $6.0\overline{)2.4}$

10.
```
   0.8
−  0.3
```

11.
```
   0.8
+  0.3
```

12.
```
   0.8
×  0.3
```

13.
```
   1.8
−  2.3
```

14.
```
   101.8
+    2.3
```

15.
```
   12.790
−   3.418
```

16.
```
    3.418
+  12.790
```

17.
```
   3.50
×  8.30
```

18. $0.820\overline{)3.608}$

19.
```
   34.10
×  70.20
```

20.
```
   34.100
×   7.020
```

21. $7.80\overline{)44.46}$

22. $0.177\overline{)47.082}$

23.
```
   6.40
−  0.058
```

24.
```
   6.40
+  0.058
```

25.
```
   7.100
×  0.470
```

26.
```
   7.31
−  1.53
```

27.
```
   7.31
+  1.53
```

Chapter 5

Algebra

Chapter Introduction

What Is Algebra?

Mathematics can be defined as the science of patterns and relationships. To teach mathematics and to learn mathematics, one must be able to determine a pattern and to define that pattern in some symbolic way using a variable or variables. Arithmetic, which includes patterns, the place value numeration system, and the properties and algorithms of addition, subtraction, multiplication, and division, relies on the laws of algebra. Algebra is the language of mathematics. It is a systematic, mathematical way of expressing abstract concepts—the branch of mathematics that involves the rules of addition, subtraction, multiplication, and division operations and numerical relationships.

Algebraic thinking is the ability to think in the abstract using the notion of a variable. A variable (often referred to as the unknown) has tremendous mathematical power because it represents infinitely many numbers. Algebra generalizes arithmetic so that letters representing numbers demonstrate the logical relationships between them.

Vocabulary of Algebra

Before introducing one or more of the reteaching models proposed in this chapter, you may need to review the following algebra terms with your students.

A *variable* is a symbol or letter that represents an unknown quantity: x, y, a, b.

An *algebraic expression* consists of numbers, variables, and arithmetic and grouping symbols: $(2x + 3)$, $5(a + b) - 6$.

An equation is a statement that two algebraic expressions are equal. All equations contain an equal sign: $5 + 6 = 11$, $a - 8 = 10$.

A *term* is a single expression, like $3a$ or $6x$. The variable is the a or x. The 3 and 6 are called *coefficients* of the term. Like terms ($6a + 4a = 10a$) can be combined. But $6a + 4b$ cannot be combined because they are not *like* terms.

A *coefficient* is usually a number teamed with a variable in an algebraic expression ($7x - 3xy$), also called a multiplicative factor.

A *constant term* is a number with no variables in an algebraic expression: $1.5 \times y$.

A *polynomial* is an algebraic expression that includes one or more variables and constants and uses only the operations of addition, subtraction, and multiplication.

A *monomial* is a term that is a numeral (7), a variable (x), or a product of a numeral and one or more variables.

A *binomial* is a polynomial of two terms: $5y^2 + 7$.

A *trinomial* is a polynomial of three terms: $a^2 - 2ab + b^2$.

Integers are the positive and negative whole numbers and zero.

A *factor* is a divisor (an integer) that evenly divides a number without leaving a remainder. A factor is also the coefficient (a multiplicative factor) in an expression.

A *function* is a set of rules for taking input (a number) and producing output (another number). If $\times 2$ is the function and the input is 7, the output will be 14. A function is a mathematical relationship that assigns only one element of a set to another element. The only outcome of 7×2 is 14.

A *quadratic equation* is a polynomial equation—one that includes one or more variables, in which one or more of the variables is squared. The Pythagorean theorem is a quadratic equation: $a^2 + b^2 = c^2$.

A *linear equation* is an algebraic equation in which each term is either a constant or the product of a constant and a single variable of the first power ($x = 2y$). Linear equations that have only one variable are the simplest equations to solve and relate to arithmetic: $2x + 4 = 12$.

Properties of Algebra

Algebra uses many of the same properties used in arithmetic operations (addition, subtraction, multiplication, and division). If necessary, take a few minutes to review these properties with your students.

Algebra has certain properties	
Commutative Property	The order in which two numbers are added or multiplied does not matter: $a \times b = b \times a$.
Associative Property	When three or more numbers are added or multiplied, the product is the same regardless of the grouping of the factors: $(a + b) + c = a + (b + c)$ and $(a \times b) \times c = a \times (b \times c)$.
Identity or Unity Property	Any number or variable multiplied by 1 is itself: $(1 \times a = a)$.
Zero Property	Anything multiplied by 0 is 0: $(0 \times a = 0)$.
Distributive Property of Multiplication Over Addition	The Distributive Property allows you to multiply a sum by multiplying each addend separately and then adding the products: $a(b + c) = ab + ac$.
Addition Property of Equality	If the same number is added to equal numbers, the sums are equal: if $a = b$, then $a + c = b + c$.
Subtraction Property of Equality	If the same number is subtracted from equal numbers, the differences are equal: if $a = b$, then $a - c = b - c$.
Transitive Property	If $a = b$ and $b = c$, then $a = c$.
Closure Property	If a is an element of set S and b is an element of set S, then a operation b is an element of set S and is unique.

Expectations for Middle School

For many years, formal algebra studies were expected to begin in ninth grade, with advanced students taking algebra in eighth grade. Due to advances in technology, understanding algebra and algebraic thinking is now expected of all students, including low-performing students. Since research demonstrates that students who study algebra in middle school are more likely to be successful in school and attend college, there has been a movement for all students to take algebra in middle school. If algebra is an extension of the foundation laid in elementary school arithmetic—learning math facts, properties, and algorithms—the transition should not be abrupt.

Common Problems Students Have with Algebra

Students learn about algebra in stages and can get stuck in any stage. The basic stages of understanding algebra and the difficulties associated with those stages are as follows.

Recognizing and Extending Patterns Recognizes a simple pattern and can extend and duplicate a repeating pattern.	*Unless students recognize patterns, they will not be able to see the pattern of relationships between numbers.*
Using Quantity Symbols Uses < and > and = symbols to relate numbers.	*If students cannot use symbols and recognize quantity relationships, they cannot work with algebraic equations.*
Identifying Functions Translates patterns to numeric equations by deducing a function when given a table of inputs and outputs.	*If students can't identify a function when they are presented with the pattern of inputs and outputs, they cannot generalize number relationships necessary for algebra.*
Calculating Unknown Values Finds unknown values in equations.	*Unless students can calculate numbers to find unknown values, they cannot work with equations.*
Relational Thinking Recognizes and uses patterns to compare two sides of a number sentence with reasoning.	*If students don't understand how to tell if two sides of an equation are equal, they cannot begin to simplify equations.*
Recognizing Simple Linear Functions Creates graphs of functions presented in input/output tables, such as tables of rates and ratios.	*Unless students recognize the linear relationships in a simple linear equation, they will not be able to build on that in working with more complicated equations.*

Relational Thinking with Addition, Subtraction, Multiplication, and Division Recognizes and uses patterns that involve arithmetic, including use of the Commutative, Associative, and Distributive Properties.	*Unless students are proficient in using the properties of arithmetic, they will not be able to translate them to working with algebraic equations.*
Identifying Variables Represents the idea of an unknown quantity or variable as a letter or symbol in an expression or equation.	*Until students can substitute letters for numbers, they will not be able to generalize number relationships in algebra.*
Calculating Exponents Recognizes and calculates simple powers of whole numbers.	*Until students understand and can find the value of exponents, they will not be able to work with algebraic expressions and equations.*
Factoring Creates factor sets and identifies the highest and lowest common factors.	*Unless students recognize how to factor numbers, they will not be able to simplify equations.*
Simplifying Fractions Simplifies fractions and finds equivalent fractions.	*Until students can simplify fractions, they will not be able to simplify equations.*
Calculating Variables Translates from verbal description to algebraic representation.	*If students cannot identify variables and solve equations with variables, they will not be able to progress in algebra.*
Representing Linear Functions Represents and constructs a table of values and graphs for linear functions.	*Unless students understand how to represent and graph linear functions, they will not be able to progress in algebra.*
Simplifying and Solving Linear Equations Simplifies and solves linear equations by elimination, transformation, or substitution.	*Unless students are able to simplify and solve linear equations, they will not be able to work with quadratic equations and develop algebraic thinking.*
Simplifying and Solving Quadratic Equations Simplifies and solves quadratic equations by elimination or substitution.	*Unless students are able to simplify and solve quadratic equations, they will not be able to progress in algebra.*

What Research Says

Research has identified the following issues relating to the difficulty of developing algebraic thinking.[1]

- Transitioning from arithmetic to algebra focuses on algebraic expressions and equations and requires thinking that is different from that used in arithmetic. Elementary arithmetic focuses on finding answers rather than understanding relationships among numbers. When struggling students see an equal sign, they interpret it as a signal to compute rather than to evaluate equivalence. Algebra calls for an analysis of the relationships among numbers, not the answer. Students who are stuck in the arithmetic mode have trouble making this transition.

- Algebra also requires the consideration of the idea that a variable can be an infinite set of values, since variables are demonstrating the relationships, not the values of numbers. Students who are looking for the right answer struggle with this transition.

- Rule-based instructional approaches without opportunities to create meaning or learn when to use rules and formulas lead to poor retention and understanding of algebra.

- Elementary students who are perceived as proficient in math are typically those who have mastered the math facts and can add, subtract, multiply, and divide accurately and quickly. Algebra requires students to think flexibly about number relationships and reason about operations to simplify equations, not just to become more proficient at procedures. Many students have a great deal of difficulty switching from a memorization and rote mode to a reasoning mode.

- Even those students who are successful in finding equivalent fractions may struggle with algebra, shifting the concept of equivalence from number to expressions.

---

[1] See National Research Council, *Adding It Up: Helping Children Learn Mathematics* (2001), for a more complete summary.

Models for Reteaching Algebra Concepts

Following are five models that teachers can use to reteach concepts at different stages of understanding algebra. These models provide alternative introductions or reteaching methods to textbook instruction. Several of the models use manipulatives to help demonstrate the relationships among numbers.

Using manipulatives not only helps students visualize and manipulate relationships instead of simply trying to memorize formulas, but it also produces enormous benefits for a better understanding of algebraic ideas.

Model:

Addition, Subtraction, Multiplication, and Division with Two-Color Counters

Recognizes and uses patterns that involve arithmetic, including use of the Commutative, Associative, and Distributive Properties.

Model adding integers to look for patterns.

One of the best manipulatives you can use to model the study of integers is two-color counters. Although available in a variety of colors, this chapter uses the red and yellow counters. They are two-sided, with red on one side and yellow on the other.

For this chapter, yellow will represent positive numbers and red will represent negative numbers. The opposite of a yellow counter (positive) is a red counter (negative).

To model a positive three (+3), simply show three yellow counters. To model a negative three (–3), show three red counters. The expressions $a - 1$ and $a + 1$ (one red and one yellow counter) cancel each other out. In other words, 1 red + 1 yellow = 0.

Model the following problems with two-color counters and look for patterns in adding negative and positive integers.

Examples

1. $^+3 + ^+4 = ^+7$ **2.** $^-3 + ^-4 = ^-7$

3. $^+6 + ^+2 = ^+8$ **4.** $^-2 + ^-5 = ^-7$

5. $^+8 + ^+3 = ^+11$ **6.** $^-7 + ^-3 = ^-10$

7. $^+4 + ^+5 = ^+9$ **8.** $^-8 + ^-1 = ^-9$

9. $^+2 + ^+9 = ^+11$ **10.** $^-6 + ^-2 = ^-8$

Ask students: Do you see a pattern? Can you state the pattern in words?

Next, model the following using two-color counters.

Examples

11. $^+4 + ^-3 = ^+1$ **12.** $^+6 + ^-9 = ^-3$

13. $^+3 + ^-4 = ^-1$ **14.** $^+2 + ^-3 = ^-1$

15. $^+5 + ^-3 = ^+2$ **16.** $^+3 + ^-3 = 0$

Ask again: Do you see a pattern? Can you state the pattern in words?

The patterns observed above provide us with the following generalizations:

A. If you add a positive number to a positive number, the result is the positive sum of the numbers.

B. If you add a negative number to a negative number, the result is the sum of the two numbers, and it will be negative.

C. If you add a negative number to a positive number, the result will be the difference between the two numbers, and the sign will be that of the greater number.

Have students practice adding positive and negative integers.

> ### Practice
>
> Solve the following using two-color counters.
>
> **1.** $^+3 + {}^+5 =$ **2.** $^+3 + {}^-5 =$
>
> **3.** $^-3 + {}^+5 =$ **4.** $^-3 + {}^-5 =$
>
> **5.** $^+2 + {}^+4 =$ **6.** $^-2 + {}^-4 =$
>
> **7.** $^-2 + {}^+4 =$

Model subtracting integers to look for patterns.

Use the two-color counters to model these equations. Have students look for patterns when subtracting negative and positive integers.

a. $^+6 - {}^+3 = {}^+3$

If we have 6 yellow counters and we take away 3 yellow counters, then we will have 3 yellow counters left, which is the difference between the two numbers.

b. $^-6 - {}^-3 = {}^-3$

If we have 6 red counters and we take away 3 red counters, then we will have 3 red counters left.

Now, let's look at another example using manipulatives:

c. $(^+6) - (^-3)$

Suppose we have 6 yellow counters ($^+6$) and want to take away 3 red counters ($^-3$), but we don't have any red counters to take away. We could add 3 red counters and 3 yellow counters instead, which in effect would be adding zero [$^+3 + {}^-3 = 0$ and $a + 0 = a$].

Result: $^+6 + (^+3 + {}^-3) - {}^-3 = {}^+6 + {}^+3 + (^-3 - {}^-3) = {}^+6 + {}^+3 + 0 = {}^+9$

Try several more of these with your students.

Here's the rule or algorithm: To subtract one integer from another, simply add the opposite. $A - B = A + {}^-B$.

There is really no such thing as subtraction in algebra. Subtraction in algebra is just adding the opposite.

Apply this new rule to:

    a. ${}^+6 - {}^+3 = {}^+6 + {}^-3 = {}^+3$

    b. ${}^-6 - {}^-3 = {}^-6 + {}^-({}^-3) = {}^-6 + {}^+3 = {}^-3$

Have students practice subtracting positive and negative integers.

Practice	
1. ${}^+4 - {}^-5 =$	**2.** ${}^-3 - {}^-2 =$
3. ${}^-3 - {}^+2 =$	**4.** ${}^-4 - {}^+2 =$
5. ${}^-3 - {}^+2 =$	**6.** ${}^+4 - {}^-2 =$
7. ${}^-3 - {}^-2 =$	**8.** ${}^-4 - {}^-2 =$
9. ${}^+3 - {}^-2 =$	**10.** ${}^+4 - {}^-2 =$

Model multiplying and dividing integers to look for patterns.

Two-color counters can help to demonstrate and examine the multiplication and division of integers. Let's look first at multiplication, which is sometimes defined as repeated addition.

First, let's look at the five components of **${}^+2 \times {}^+3$**:

  1. A number (3)

  2. The sign of that number + or – (+)

  3. The multiplication sign (×)

  4. The number of sets (2)

  5. The sign of the number of sets + or – (+)

Then, let's look at how to solve the problem:

1. Get three counters.

2. They are + (positive), so they are yellow.

3. We want to make sets.

4. We want to make 2 sets of 3 yellow, which equals 6 yellow.

5. Multiplying a positive number by a positive number results in a positive number.

6. The answer is 6 yellow, or $^+6$.

Now, model these:

$$^+2 \times {}^-3 = ?$$

1. Get three counters.

2. They are – (negative), so they are red.

3. We want to make sets.

4. We want to make 2 sets of 3 red, which equals 6 red.

5. Multiplying a negative number by a positive number results in a negative number.

6. The answer is 6 red, or $^-6$.

$$^-2 \times {}^-3 = ?$$

1. Get three counters.

2. They are red, negative.

3. We want to make sets.

4. We want to make 2 sets of 3 red, which equal 6 red or $^-6$.

5. Negative ($^-2$) means take the opposite of $^-6$, so it equals $^+6$. Multiplying two negative numbers results in a positive number.

6. The answer is $^+6$.

$^-2 \times {}^+3 = ?$

1. Get three counters.

2. They are yellow, positive.

3. We want to make sets.

4. We want to make 2 sets of 3 yellow, which equal 6 yellow or $^+6$.

5. Negative ($^-2$) means to take the opposite of $^+6$. Multiplying a negative and a positive number results in a negative number.

6. The answer is $^-6$.

Ask students: Do you see the pattern?

When you multiply *like* signs, the answer is always *positive*.

When you multiply *unlike* signs, the answer is always *negative*.

Division is the opposite of multiplication, but it uses the same pattern.

When you divide *like* signs, the answer is always *positive*.

When you divide *unlike* signs, the answer is always *negative*.

Have students practice multiplying and dividing integers.

Practice
Solve each of these problems using two-color counters.

1. $4 \times 4 =$

2. $^-4 \times 4 =$

3. $^-4 \times {}^-4 =$

4. $^-4 \div 4 =$

5. $^-4 \div {}^-4 =$

6. $4 \div {}^-4 =$

7. $4 \div 4 =$

Model:
Exponents (moving from patterns to rules)

Recognizes and calculates simple powers of whole numbers.

Introduce exponents.

Variables generalize arithmetic. Algebra makes arithmetic infinite because a variable can stand for infinitely many numbers. If arithmetic expresses a math fact, algebra expresses the rule that applies the same relationship to any number. For example, if a equals any whole number, then $a + 3$ equals three more than any whole number.

Let's look for the pattern in these equations and generalize a set of rules algebraically to state the rule.

a. $3^2 \times 3^4 = 3 \cdot 3 \cdot 3 \cdot 3 \cdot 3 \cdot 3 = 3^6$

b. $5^3 \times 5^2 = 5 \cdot 5 \cdot 5 \cdot 5 \cdot 5 = 5^5$

c. $9^2 \times 9^2 = 9 \cdot 9 \cdot 9 \cdot 9 = 9^4$

What's the pattern? Answer: $a^x \times a^y = a^{x+y}$

Rule When the base is the same, you add exponents when multiplying.

d. $\dfrac{3^4}{3^2} = \dfrac{3 \cdot 3 \cdot 3 \cdot 3}{3 \cdot 3} = 3 \times 3 = 3^2$

e. $\dfrac{6^8}{6^4} = \dfrac{6 \cdot 6 \cdot 6 \cdot 6 \cdot 6 \cdot 6 \cdot 6 \cdot 6}{6 \cdot 6 \cdot 6 \cdot 6} = 6^4$

f. $\dfrac{2^3}{2^2} = \dfrac{2 \cdot 2 \cdot 2}{2 \cdot 2} = 2 = 2^1$

What's the pattern here? Answer: $\dfrac{a^x}{a^y} = a^{x-y}$

Rule When the base is the same, you subtract exponents when dividing.

Here are some examples that are a little different.

g. $\dfrac{3^2}{3^3} = \dfrac{3 \cdot 3}{3 \cdot 3 \cdot 3} = \dfrac{1}{3}$

h. $\dfrac{4}{4^2} = \dfrac{4}{4 \cdot 4} = \dfrac{1}{4}$

What's the pattern for these?

Using the rule for examples **d–f**, $\frac{a^x}{a^y} = a^{x-y}$, let's see what we would get:

g. $3^{2-3} = 3^{-1}$, which must equal $\frac{1}{3}$.

h. $4^{1-2} = 4^{-1}$, which must equal $\frac{1}{4}$.

Now, how can we write the rule algebraically?

$\frac{a^x}{a^y} = a^{x-y}$ and $a^{-x} = \frac{1}{a^x}$

Here's another question. What is $\frac{a^3}{a^3}$?

$a^{3-3} = a^0$. What's the zero power?

To find out what the zero power is, let's look at an arithmetic example: $\frac{6^3}{6^3}$.

$\frac{6 \cdot 6 \cdot 6}{6 \cdot 6 \cdot 6} = 1$, so $\frac{6^3}{6^3} = 6^0$, which must equal 1, too.

Rule Any number to the zero power must equal 1.

Have students practice with exponents.

Practice
Solve these equations.

1. $\frac{4^2}{4^3} =$ **2.** $\frac{4^3}{4^2} =$

3. $\frac{4^3}{4^3} =$ **4.** $\frac{a^2}{a^3} =$

5. $\frac{a^3}{a^3} =$

Model:	Represents the idea of an unknown quantity or variable as a letter
Variables	or symbol in an expression or equation.

Have students practice describing patterns in words and symbols.

Practice

Complete the following exercises.

1. **a.** 1, 3, 4, 7, 11, 18, ___, ___, ____, …

b. Describe the pattern in words.

c. Describe the pattern in symbols.

2. **a.** 1, 2, 4, 8, 16, 32, ___, ___, ___, …

b. Describe the pattern in words.

c. Describe the pattern in symbols.

3. **a.**

x	y
3	10
0	7
−8	−1
−13	?

b.

x	y
1	−6
−2	−9
3	−4
0	?

c.

x	y
4	13
−5	16
2	7
3	?

d. For a–c, describe the pattern in words.

e. For a–c, describe the pattern in symbols.

Have students practice translating words into symbols and symbols into words.

Practice

1. Translate each sentence into numbers and symbols.

a. Three times a number squared

b. Eight more than three times a number

c. Eleven times the sum of a number plus six

d. Four more than the sum of a number decreased by six

e. The sum of a number and that number increased by four equals 24

2. Translate these equations into words.

    a. $x + 9 = 16$ **b.** $3(m + 2) - 6 = 24$

    c. $7 = 4x + 2$ **d.** $3x - 7 = 13$

Demonstrate building equations with variables.

Here is an example of a good problem to solve using algebra.

$$\ldots 12, 10, 8, 6, \underline{\hphantom{xx}}, \underline{\hphantom{xx}}, \underline{\hphantom{xx}},$$

What comes next? To think algebraically (to find a pattern), we need some data. Arrange the data to find a pattern.

[N] Number in Sequence	[P] Position
12	1
10	2
8	3
6	4
4	5

Let F represent the first number in the sequence because it does not have to be 12. But if 12 is the starting point, we could write F as $12 - 0$, the second number as $12 - 2 \times 1$, the third number as $12 - 2 \times 2$, the fourth number as $12 - 2 \times 3$, and so on. Here is an easier way to spot the pattern.

N	P
$12 = (12 - 0)$	1
$10 = (12 - 2 \times 1)$	2
$8 = (12 - 2 \times 2)$	3
$6 = (12 - 2 \times 3)$	4
$4 = (12 - 2 \times 4)$	5

$N = \text{(first number)} - 2(P - 1)$ $N = F - 2(P - 1)$

Therefore, to find the 100th term, just plug the numbers into our "formula". That is, P is the position and N is the number in the sequence we are looking for.

$N = 12 - 2(100 - 1) =$ $N = 12 - 2(99) = {}^{-}186$

Have students practice writing equations with variables to represent patterns.

Look at the sequence in the preceding example as it increases to 14, 16, 18, and higher. What would change in the formula $N = F - 2(P - 1)$?

Model:

Linear Equations with "Cup" Algebra

Simplifies and solves linear equations by eliminating or substituting.

Introduce simplifying expressions.

It is usually a good idea to simplify expressions before solving an equation. The process is similar to simplifying fractions to find the lowest common denominator (say, $\frac{4}{8}$ to $\frac{1}{2}$).

For example:

$$^-8x + 4x = {}^-4x$$

$$\frac{1}{2}y + \frac{2}{3}y - \frac{1}{6}y = \frac{3}{6}y + \frac{4}{6}y - \frac{1}{6}y = \frac{6}{6}y = 1y$$

Remember, you can only simplify *like* terms.

Have students practice simplifying expressions.

Practice

Simplify.

1. $2x^2 + 4x^2$

2. $\frac{1}{2} + \frac{3}{2}x - x + \frac{3}{2}$

3. $4(x + 2) - 3(2x + 5)$

4. $20x - 4x$

5. $8 + 3x - 4 + 2x$

6. $\dfrac{\frac{1}{2}c}{2c}$

7. $\frac{3}{4}y - \frac{2}{3}y + 4$

8. $2x + 5x - 6x + x$

9. $3x + 6y + 3x + 9y$

10. $2x + 3x - 2y + 3y$

Introduce "cup" algebra for solving linear equations.

Solving algebraic equations is really a very simple procedure, but students often become confused about how to manipulate the symbols. They rarely understand what they are doing unless given a concrete introduction. "Cup" algebra is one method that may help students understand what they are doing as they solve equations.

First, let's define the manipulatives in terms of the abstract ideas they represent or model. We will use two manipulative materials: a cup (any size) or portion cups and two-color counters.

- A yellow two-color counter, represented by a white circle in the following illustrations, will represent $^+1$.

Yellow

- A red two-color counter, represented by a shaded circle, will represent $^-1$.

Red

- An upright cup will represent x, or ^+x.

- An upside-down cup will represent ^-x, or opposite x.

Here is something to keep in mind: An equation is like a seesaw. For the equation to be balanced, whatever is on one side must be equivalent to what's on the other. In other words, you can do anything you want to one side if you do the same on the other. Remember that if $a = b$, then $a + c = b + c$, and $a + {}^-a = 0$.

Yellow   Red

$\bigcirc$ + $\bullet$ = 0

Demonstrate "cup" algebra using these examples.

Example 1 $x + 3 = 7$

$x + (^+3) = ^+7$

$x + ^+3 + ^-3 = ^+7 + ^-3$

$x + 0 = ^+4$

$x = 4$

Example 2 $2x - 1 = 5$

$2x - 1 = 5$

$2x - 1 + 1 = 5 + 1$

$2x = 6$

$x = 3$

Example 3 $2(x + 1) = x + 6$

$2(x + 1) = x + 6$

$2x + 2 = x + 6$

$2x + 2 - 2 = x + 6 - 2$

$2x = x + 4$

$2x - x = x - x + 4$

$x = 4$

Have students practice solving linear equations.

Practice
Solve for x with or without counters and cups.

1. $3(x + 3) = 2x + 1$ **2.** $x + 1 + 3x = x + 2(x - 2)$

3. $-x + 3 = 2x - 5$ **4.** $9 + x = 2(3 - x)$

Once students have grasped the idea of "cup" algebra, it's a good exercise to have them make up their own problems using cups and two-color counters.

Introduce solving linear equations using the Addition and Subtraction Properties of Equality.

As we learned in the previous section, when solving equations, there is one basic thing to remember: You can do anything you want to an equation (except divide by zero) *as long as you do the same thing to both sides.*

For example:

If $x + 6 = 3$, subtract 6 from both sides:

$$x + 6 - 6 = 3 - 6$$

$$x = {}^-3$$

If $5(x + 4) = 2x + 2$, simplify first:

$$5x + 20 = 2x + 2$$

Then subtract 20 and $2x$ from both sides:

$$5x + 20 - 20 - 2x = 2x + 2 - 20 - 2x$$

$$3x = 2 - 20$$

$$3x = {}^-18$$

$$x = {}^-6$$

We subtracted 20 from both sides and also subtracted $2x$ from both sides to simplify the equation to $3x = {}^-18$, and then divided both sides by 3 to get $x = {}^-6$.

The preceding examples illustrate a basic rule: to solve an equation, we need to get all of the variables (x's) on one side of the equation and all of the numbers (or things we are not solving) on the other.

Have students practice solving linear equations.

Practice

Solve the following equations using the Addition and Subtraction Properties of Equality.

1. $5x = 15$

2. $3x = -8$

3. $-8x = 0$

4. $36x = 0$

5. $\frac{1}{2}y = \frac{2}{3}$

6. $\frac{3}{4}x = \frac{6}{8}$

7. $25x = 5$

8. $2(4x - 4) + 1 = 3(x + 2)$

9. $3(2y - 6) + 5y = 2(y + 5)$

10. $3x = 3(x - 1) + x$

11. $5x = 3(x - 2)$

12. $4(x - 2) + 2(2x + 4) = 10$

13. $5y = 15 - 2(y - 7)$

14. $y - (3y + 2) = 15 + 2y$

15. $x - 5 = 5x - 3$

16. $4(x - 3) = 5 + (x + 4)$

17. $-13y + 2 = 22 - 2y$

18. $2(4x - 1) - 2(x + 2) = 2(x - 1)$

19. $3y + 2 = y + 5$

20. $\frac{1}{3}x + \frac{1}{4} = \frac{1}{6} + x$

Model:
Polynomials with Algebra Tiles

Simplifies and solves quadratic equations by eliminating or substituting.

Develop an understanding of polynomials and the principles of algebra using manipulatives.

Educators have long echoed the philosophy of building understanding from the concrete to the semi-concrete to the abstract. While this has been recognized for years as the correct way to teach mathematics, it has only recently been incorporated into mathematics instruction. The farther one moves up the mathematical ladder, the less manipulatives are available to help make concepts concrete. Furthermore, some concepts are just too abstract for concreteness.

Fortunately, a few years ago a number of algebraic manipulatives came into the marketplace that greatly assist educators in teaching some important algebraic concepts and helping students better understand those ideas. Math tiles, algebra tiles, and other similar products became available to help learners better understand polynomials. It is absolutely vital that students have hands-on experiences with algebra tiles/blocks when learning algebraic concepts.

Using algebra manipulatives, students can understand linear equations that have no exponents and quadratic equations that may have exponents to the power of 2.

To model a polynomial with manipulatives, let's start with algebra tiles. Most algebra tiles come in the following three forms:

| Unit block | Rod | Square |

Unit blocks are yellow on one side and red on the other. The rods are green on one side and red on the other. The squares are blue on one side and red on the other.

It is important to recall the mathematical principles you will use as you work with the tiles:

Principle 1. $A + 0 = A$

Principle 2. $A + {}^-A = 0$

Principle 3. $A - B = A + {}^-B$

Note: Algebra tiles or blocks are purposely made so that one shape cannot be made to cover an unlike shape. That is, the tiles are *noncommensurable:* because one variable is not necessarily a multiple of the other, an x^2 tile cannot be covered exactly by x tiles, nor can x tiles be covered exactly by unit tiles.

In algebra, a variable such as x represents some unknown quantity, usually a number. You will notice that only one variable, x, is used in these exercises rather than using both x and y variables. Working with one variable will simplify students' concept development. If a question about a second variable arises, make the value of the second variable, y, equal 1, so students can focus on the algebraic relationships and not worry about tracking two variables.

First, define the three tile shapes as shown following:

Yellow unit = ☐ = 1   The unit tile is a square whose sides are each 1, so the area equals 1 ($1 \times 1 = 1$).

Green rod = [diagram, side labeled x and 1]   One side of the rod is the same length as the unit, 1, and the second side is some unknown; let's call it x.

Blue square = [diagram, sides labeled x and x]   Each side of the blue square is the same length as the rod, x. So the area is $x \cdot x$, or x^2.

To change any number represented by a tile to a negative number, turn the tile over so that it is a different color. Make sure that all negative numbers are represented by one color and all positives are represented by another color.

Now, let's review our definition and principles.

Yellow unit = +1   Red unit = −1

Green rod = +x Red rod = −x

Blue square = +x^2 Red square = −x^2

Yellow unit (+1) + red unit (−1) = 0

Green rod (x) + red rod (−x) = 0

Blue square (x^2) + red square (−x^2) = 0

Model polynomials with algebra tiles.

In the illustrations that follow, the shaded figures represent the red (negative) tiles.

$x^2 + 1$

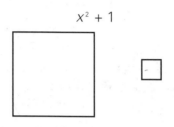

$2x + 3$

$x^2 + 3x - 4$

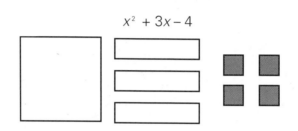

Have students practice modeling polynomials.

Practice
Model the following polynomials using algebra tiles.

1. $2x^2$ **2.** $3x$

3. $x^2 + 2x$ **4.** $2x^2 + 3x + 1$

5. $x^2 + 2x + 3$ **6.** $x^2 - x - 1$

7. $2x^2 - 2x + 3$ **8.** $3x^2 + 2x - 4$

9. $x^2 + 2x - 5$ **10.** $x^2 - 4$

Model adding polynomials.

To model the addition of polynomials, simply put together the algebra tiles indicated by the given polynomial. Here are some examples:

a. $x^2 + 2x + 1$

$+\ x^2 + \ x + 3$

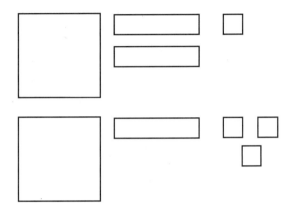

The answer is $2x^2 + 3x + 4$.

b. $2x^2 + x - 3$

$+\ \ x^2 - x + 2$

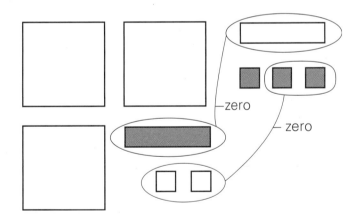

The answer is $3x^2 - 1$.

c. $3x^2 - 4x + 5$
$+ \quad\quad 3x - 6$

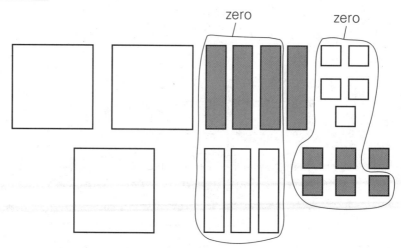

zero   zero

The answer is $3x^2 - x - 1$.

Have students practice adding polynomials without using algebra tiles.

Practice

Solve these addition problems. Look for patterns that let you solve the problems without using manipulatives.

1. $\quad 3x^2 - 2x + 6$
$\quad + \quad x^2 + x - 3$

2. $\quad 2x^2 - 3x - 1$
$\quad + \quad x^2 + x - 2$

3. $\quad x^2 + 5x - 4$
$\quad + x^2 + \quad x + 5$

4. $\quad x^2 + 5x - 4$
$\quad + x^2 - 4x + 5$

5. $\quad x^2 \quad\quad - 2$
$\quad + \quad x^2 + x - 3$

6. $\quad x^2 - 2x - 1$
$\quad + x^2 + \quad x + 2$

7. $\quad x^2 + 4 \ - 5x$
$\quad + \quad x^2 + 2x - 3$

8. $\quad 2x^2 - 1 \ - 3x$
$\quad + \quad 5x + x^2 - 4$

Model the subtraction of polynomials.

Subtracting polynomials sometimes provides an extra challenge. Here are a few examples:

a. $2x^2 + x + 3$

$-\ (x^2 + x + 2)$

$2x^2 + x + 3$:

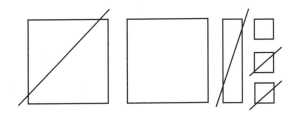

Take away: x^2 (1 blue square), x (1 green rod), and 2 yellow units:

The remainder is $x^2 + 1$.

So the answer is $(2x^2 + x + 3) - (x^2 + x + 2) = x^2 + 1$.

b. $3x^2 - 5x + 2$
$- 2x - 4$

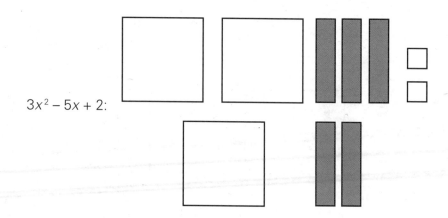

$3x^2 - 5x + 2$:

That is, take away 2 green rods and 4 yellow units.

But we don't have 2 green rods and 4 yellow units. There are no green rods or red units to subtract. But if we recall the principles, we can solve the problem. All we need to do is add 4 red and 4 yellow units, or the equivalent of zero, and then take away 4 red units. Similarly, add zero in the form of 2 red rods and 2 green rods:

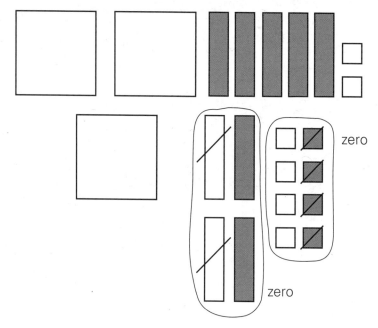

The polynomial remaining is $3x^2 - 7x + 6$.

Therefore, the answer is $(3x^2 - 5x + 2) - (2x - 4) = 3x^2 - 7x + 6$.

This same procedure can be used anytime it becomes necessary. However, there is another method that some people believe is better. This second method uses the principle $A - B = A + {}^-B$ and is demonstrated by the following example.

$$2x^2 + x - 3$$
$$-\quad (x^2 - x + 4)$$

$2x^2 + x - 3$:

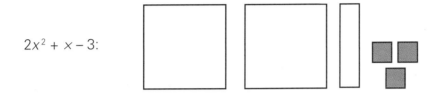

Subtract $(x^2 - x + 4)$:

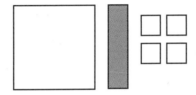

But if $A - B = A + {}^-B$, this means we can subtract simply by taking the opposite of $(x^2 - x + 4)$.

The opposite of $x^2 - x + 4$ is ${}^-x^2 + x - 4$, as shown here:

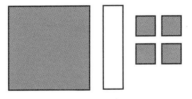

And now we can add, as previously modeled.

$$2x^2 + x - 3$$
$$+\quad {}^-x^2 + x - 4$$
$$\overline{\quad x^2 + 2x - 7\quad}$$

The answer is $x^2 + 2x - 7$.

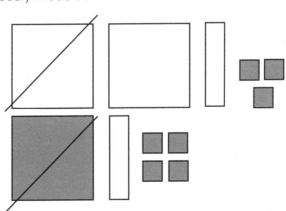

Have students practice subtracting polynomials.

Practice

Solve the following. Can you do these without using algebra tiles?

1. $(x^2 - x + 3)$
  $+ (x^2 - 2x + 1)$

2. $(x^2 + 2x - 3)$
  $+ (2x^2 + x - 1)$

3. $(2x^2 + x - 1)$
  $+ (x^2 - 3x + 4)$

4. $(2x^2 + 3x + 4)$
  $- (x^2 - 4x - 1)$

5. $(x^2 - x - 1)$
  $- (x^2 + x - 1)$

6. $(3x^2 - 4)$
  $- (2x^2 + x - 1)$

7. $(2x^2 - x + 3)$
  $- (x^2 + x - 4)$

Model the multiplication of polynomials.

The model using algebra tiles to multiply polynomials builds on the idea that areas can represent multiplication. For example, if we place 3 rods to form a rectangle, the dimensions of the rectangle are 3 by x.

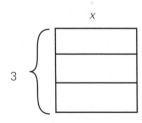

Therefore, the area would be $3x$.

Model the following with algebra tiles and find the area.

The top side of this figure has a length of $(x + 2)$, so the area would be $4(x + 2)$, or $4x + 8$.

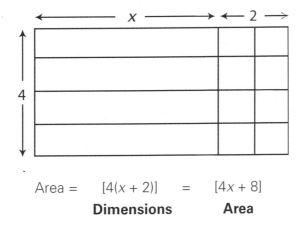

Area = $[4(x + 2)]$ = $[4x + 8]$

Dimensions   **Area**

In the next figure, the top side has a length of $(x + 2)$, and the vertical side has a length of $(x + 1)$:

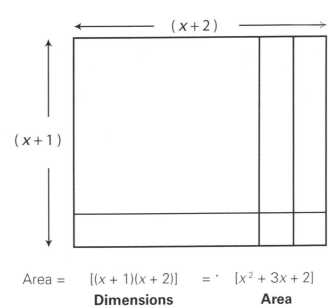

Area = $[(x + 1)(x + 2)]$ = ` $[x^2 + 3x + 2]$

Dimensions   **Area**

Now, model this example with algebra tiles: $2x(x + 5)$.

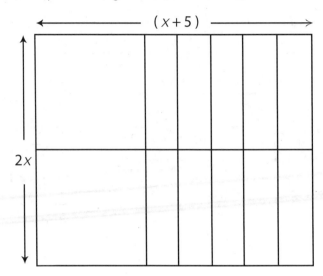

$$\text{Area} = 2x(x + 5) = 2x^2 + 10x$$

Have students practice multiplying polynomials.

Practice

Show the following using algebra tiles. See if you can discover a pattern to solve the problems without using algebra tiles.

1. $x(x + 1) =$ **2.** $(x + 1)(x + 3) =$

3. $(2x + 1)(x + 3) =$ **4.** $(x + 2)(x + 2) =$

5. $(x + 1)(x + 4) =$ **6.** $(x + 3)(2x + 2) =$

Model the multiplication of polynomials with negative numbers.

The preceding problems present no real difficulties because none of the dimensions involve negative numbers. The following models will examine how to handle those special cases.

If we begin with a square whose dimensions are x by x, the area is x^2.

When a rod is subtracted, the dimensions of the figure become x by $x - 1$. Therefore, the area would be $x(x - 1) = x^2 - x$.

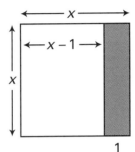

But remember, the algebra tiles are noncommensurable, so instead of placing tiles on top of other tiles, we will place them next to each other:

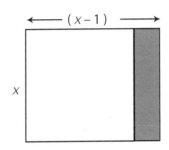

Area = $x^2 - x$

Let's look at $(x + 1)(x - 2)$ next.

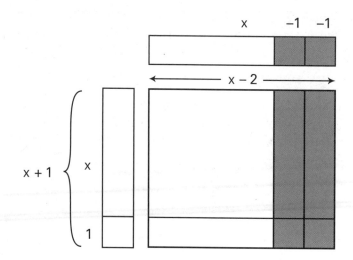

The shaded rods and units signify that we are subtracting from the whole rectangle. The answer is $(x + 1)(x - 2) = x^2 - 2x + x - 2 = x^2 - x - 2$.

Now, let's examine the model for $(x - 1)(x - 2)$.

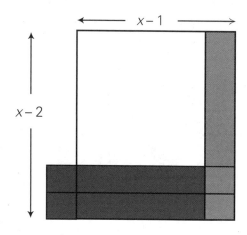

As the model shows, subtracting the rods would actually duplicate a portion of the figure—the part sticking out of the figure. Therefore, $(x - 1)(x - 2) = x^2 - 3x + 2$.

Have students practice multiplying polynomials with negative numbers.

Practice

Model the following with algebra tiles.

1. $(x - 3)(x + 3)$

2. $(x - 2)(x - 2)$

3. $(3x - 4)(x + 2)$

4. $(x - 1)(x - 3)$

5. $(x - 4)(x - 3)$

6. $(x + 3)(2x - 1)$

Model dividing and factoring trinomials.

When we look at the models for multiplication, division, and factoring, we can see how these three operations are closely related. For example, suppose we model $x^2 + 5x + 6$:

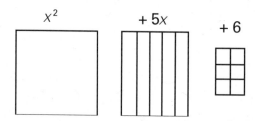

How can we arrange all these pieces in a rectangle so that one dimension is $x + 2$?

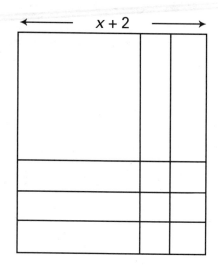

Looking at this figure, it is obvious that the other dimension must be $x + 3$. Therefore,

$$\frac{(x^2 + 5x + 6)}{(x + 2)} = x + 3$$

Now, model: $\dfrac{(x^2 - x - 2)}{(x - 2)}$

These are the parts that make up $x^2 - x - 2$:

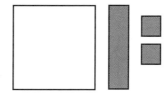

This model shows a special arrangement of $x^2 - x - 2$. To complete the rectangle, fill up the vacancies by adding x and $-x$:

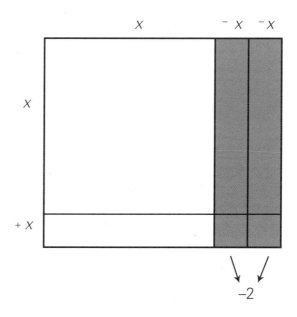

If we use the $A + {}^{-}A = 0$ principle, it can be factored as $(x - 2)(x + 1)$, which are the dimensions of the figure.

Next, model finding the factors for $x^2 + 7x + 6$.

Following are the parts that make up $x^2 + 7x + 6$. The question is how do we arrange these pieces to form a rectangle? And once that is done, what are the rectangle's dimensions?

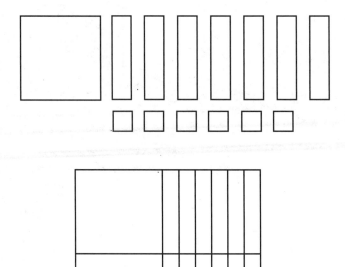

The units give us a hint about how this can be done. We could arrange the six units as 2×3 (3×2) or 1×6 (6×1). You will notice that the 2×3 arrangement will not allow all the rods to be placed to make a rectangle. The 1×6, however, does.

So the dimensions of this figure are $(x + 1)$ and $(x + 6)$.

Therefore, $x^2 + 7x + 6 = (x + 1)(x + 6)$.

Assess dividing and factoring trinomials.

Have students solve the following problems, using algebra tiles if necessary.

1. Factor $x^2 + 5x + 6$

2. Factor $x^2 - x - 2$

3. Factor $2x^2 + 5x - 3$

4. Factor $x^2 - 4$

5. Factor $4x^2 - 9$

6. Factor $x^2 - 16$

7. $\dfrac{x^2 + 4x + 3}{x + 1}$

8. $\dfrac{x^2 - 3x - 4}{x - 4}$

9. $\dfrac{x^2 + 5x + 4}{x + 1}$

10. $\dfrac{x^2 - x - 6}{x + 2}$

Model:
Mental Math

Represents, analyzes, and generalizes a variety of patterns.

Math is the science of patterns and relationships, and the following examples show how beautiful that can be. Because mathematics is such a logical system, it may not be obvious at first glance that patterns exist throughout. While these patterns are not always obvious, discerning them is a skill that math students can and should develop. Once they begin to see these patterns and relationships, they can replace calculations with mental math in many situations. Mental math is a skill unto itself, is fun to do, and can motivate students mathematically.

In the following examples, you will ask students to see if they can spot the relationships between numbers that compose a pattern, and then you will use algebra to show why these patterns exist. In fact, algebra is so powerful that it can be used to clarify the "tricks" often used in mental math.

Demonstrate patterns that occur when multiplying by eleven.

Let's look at the patterns that occur when multiplying by eleven. Here are some examples:

23	27	36	45	53	314	234
× 11	× 11	× 11	× 11	× 11	× 11	× 11
23	27	36	45	53	314	234
23	27	36	45	53	314	234
253	297	396	495	583	3,454	2,574

Do you see the pattern? Notice that in the first problem you simply bring down the "bookends" 2 and 3 in the answer 253 and then add 2 + 3 to get 5, the middle number, and that is the answer!

Does this pattern hold when multiplying a three-digit number by 11? Let's look at 314 × 11. Again, bring down the bookends 3 + 4, but instead of adding 3 + 4 to get the middle number, add 3 + 1 to get 4 and 1 + 4 to get 5. Now you have the answer: 3,454.

In the preceding examples, did you notice that the sum of the digits in the number being multiplied by 11 was always less than 10? What do you think would happen if the sum of the digits were greater than 10? Let's find out.

74	96	259	346
× 11	× 11	× 11	× 11
74	96	259	346
74	96	259	346
814	1,056	2,849	3,806

In the first problem, 74 × 11, notice what happens. If you bring down the bookends (7 and 4) and add 7 + 4 to get the middle number, you will have 7,114 as the answer. But this isn't correct. You will have to "carry over" the 1 in the 11 to get the correct answer: 814.

To review:

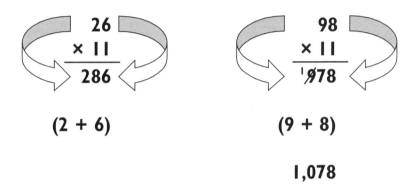

(2 + 6)   (9 + 8)

1,078

Have students practice multiplying by 11 using mental math.

Practice
Use mental math to solve these problems.

1. 34 × 11   **2.** 27 × 11   **3.** 23 × 11   **4.** 38 × 11   **5.** 425 × 11 |

Demonstrate using mental math to square numbers ending in 5.

Next, let's look at the squares of two-digit numbers ending in five and see if we can find a pattern.

15	25	35	45	55
× 15	× 25	× 35	× 45	× 55
2 25	6 25	12 25	20 25	30 25

Ask students: Do you see the pattern?

First, all of the answers end in "25." Second—and this pattern isn't so obvious—in each problem, if you increase one of the numbers in the tens place by 1 and then multiply it by the other number in the tens place, you will get the first number in the answer.

Now let's look at this pattern algebraically:

$15^2 = (10 + 5)(10 + 5) = 100 + 50 + 50 + 25 = 225$

$25^2 = (20 + 5)(20 + 5) = 400 + 100 + 100 + 25 = 625$

$35^2 = (30 + 5)(30 + 5) = 900 + 150 + 150 + 25 = 1,225$

$45^2 = (40 + 5)(40 + 5) = 1,600 + 200 + 200 + 25 = 2,025$

$55^2 = (50 + 5)(50 + 5) = 2,500 + 250 + 250 + 25 = 3,025$

$(10a + 5)$ represents any two-digit number ending in 5.

$(10a + 5)(10a + 5)$ represents any two-digit number ending in 5^2.

$(10a + 5)(10a + 5) = 100a^2 + 100a + 25 = 100a(a + 1) + 25$.

This proves what we observed in the pattern.

Have students practice squaring numbers that end in 5.

Practice

Use mental math to solve these problems.

1. 65^2

2. 75^2

3. 85^2

4. 95^2

Demonstrate how to square two-digit numbers that do not end in 5.

But what happens if the two-digit number does not have a 5 in the ones place? Let's look at the situation algebraically.

Let $(10a + b)$ represent any two-digit number, with a representing the digit in the tens place and b representing the digit in the ones place.

So $(10a + b)(10a + b)$ represents any two-digit number squared.

Therefore, $(10a + b)(10a + b) = 100a^2 + 20ab + b^2$.

Have students try using this formula to square any two-digit number. If you're thinking, "this isn't easy to do in my head," you're correct. However, with a little practice, your students might surprise you. This author has had many fifth- and sixth-grade students who could square numbers mentally using this formula faster than their teacher could do it on a handheld calculator.

Have students practice using mental math to square numbers that do not end in 5.

Practice

Use the algebraic formula $100a^2 + 20ab + b^2$ to solve these problems.

1. 34^2 **2.** 27^2 **3.** 53^2 **4.** 74^2 **5.** 87^2

Demonstrate the "one up and one down" pattern for multiplying using mental math.

Here is a pattern called "one up and one down." It can be used as a mental math shortcut to find the product of two numbers, when one factor is one less and the other factor is one more than the factors of a square.

$$8 \times 8 = 64$$

$$7 \times 9 = 63$$

$$4 \times 4 = 16$$

$$3 \times 5 = 15$$

$$15 \times 15 = 225$$

$$14 \times 16 = 224$$

$$10 \times 10 = 100$$

$$9 \times 11 = 99$$

$$50 \times 50 = 2,500$$

$$49 \times 51 = 2,499$$

Ask students: Do you see the pattern?

Does this "trick" always work? Let's use algebra to find out.

$A \times A = A^2$

$(A - 1)(A + 1) = A^2 - 1$

This proves our observation about multiplying *any* two numbers when one number is one less and the other number is one more than the factors of a square.

Have students practice using mental math to find products of numbers that fit the "one up, one down" pattern.

Practice
Use mental math to solve these problems.

1. $11 \times 13 =$ **2.** $19 \times 21 =$ **3.** $29 \times 31 =$

4. $499 \times 501 =$ **5.** $999 \times 1{,}001 =$

Demonstrate using mental math to multiply two numbers in which the tens digits are the same and the sum of the ones digits is equal to 10.

Here's another example to look at. Ask students: What do you notice?

$$
\begin{array}{r} 72 \\ \times\ 78 \\ \hline 5{,}616 \end{array}
\qquad
\begin{array}{r} 58 \\ \times\ 52 \\ \hline 3{,}016 \end{array}
\qquad
\begin{array}{r} 64 \\ \times\ 66 \\ \hline 4{,}224 \end{array}
\qquad
\begin{array}{r} 93 \\ \times\ 97 \\ \hline 9{,}021 \end{array}
$$

There are several things to notice that are true for all four examples:

1. The digits in the tens place are the same for both numbers.

2. The two digits in the ones place add up to 10.

3. The product of the numbers in the ones place gives the numbers in the tens and ones places in the answer.

4. If you increase one of the numbers in the tens place by 1 and then multiply the two numbers in the tens place, you get the first part of the answer (the numbers in the thousands and hundreds places).

In other words, take any two numbers in which the sum of the digits in the ones place is equal to 10 and the tens digits are the same—for example, 72 and 78. The product of these two numbers can be easily found using the method just described.

1. Multiply the ones digits (in this example, $2 \times 8 = 16$).

2. Increase one of the tens digits by one and multiply the other tens digit by that number (in this example, $7 \times 8 = 56$).

3. Now combine 56 and 16 to make the number 5,616.

4. And you have the answer: $72 \times 78 = 5{,}616$.

Here's why this works:

$$(10a + b)(10a + c) \quad = 100a^2 + 10ab + 10ac + bc$$

$$= 100a^2 + 10a(b + c) + bc$$

But, $b + c = 10$, so $100a^2 + 100a + bc$.

And, $100a(a + 1) + bc$ proves the "why" of the trick.

Have students practice multiplying numbers in which the tens digits are the same and the sum of the ones digits is 10.

Practice
Use mental math to solve these problems.

1.	**2.**	**3.**	**4.**	**5.**
22	34	41	57	68
× 28	× 36	× 49	× 53	× 62

Demonstrate using mental math to multiply numbers that are less than but close to 100.

This last set of problems is the most interesting and the best test of students' ability to discover a pattern. Take a look at the following:

96	93	87	89
× 95	× 96	× 95	× 91
9,120	8,928	8,265	8,099

It's pretty hard to see much in the way of a pattern, isn't it? But let's examine the first problem, 96 × 95 = 9,120, more closely.

How far from 100 is 96? The answer is 4.

How far from 100 is 95? The answer is 5.

What is 4 × 5? The answer is 20. That's interesting. Part of the answer is 20.

Here's another observation. 5 from 96 is the same as 4 from 95. Both differences are 91. Interesting again. The other part of the answer is 91.

$$96 \qquad 100 - 4 \qquad 4 \times 5 = 20$$

$$95 \qquad 100 - 5$$

$$96 - 5 = 91$$
$$95 - 4 = 91$$

Answer: 9,120

Do you see the pattern?

Let's see if this pattern holds in the second problem, $93 \times 96 = 8{,}928$.

How far from 100 is 93? The answer is 7.

How far from 100 is 96? The answer is 4.

Interesting. $7 \times 4 = 28$, a part of the answer.

Another observation: $93 - 4$ and $96 - 7$ both equal 89, the other part of the answer.

$$93 \qquad 100 - 7 \qquad 4 \times 7 = 28$$

$$\underline{\times\, 96} \qquad 100 - 4$$

$$\begin{array}{ll} 93 \quad\diagdown\;7 \\ \underline{\times\, 96} \quad\diagup\;5 \end{array} \qquad \begin{array}{l} 93 - 4 = 89 \\ 96 - 7 = 89 \end{array}$$

Answer: 8,928

Here's another:

$$87 \qquad 100 - 13 \qquad 5 \times 13 = 65$$

$$\underline{\times\, 95} \qquad 100 - 5$$

$$\begin{array}{ll} 87 \quad\diagdown\;13 \\ \underline{\times\, 95} \quad\diagup\;5 \end{array} \qquad \begin{array}{l} 87 - 5 = 82 \\ 95 - 13 = 82 \end{array}$$

Answer: 8,265

We can prove that this always works for numbers that are less than or close to 100 using algebra:

$(100 - a)(100 - b)$ represents two numbers that are less than 100 but close.

$$(100 - a)(100 - b) = 10{,}000 - 100b - 100a + ab$$

$$10{,}000 - 100(a + b) + ab$$

$$Ta$$
$$\times\,Tb$$

T = number in tens place

a, b = ones digits – 100

Here's an example:

$$96 \times 95 = (100 - 4)(100 - 5) = 10{,}000 - 500 - 400 + 20 =$$

$$10{,}000 - 900 + 20, \text{ which is } 10{,}000 - 100(a + b).$$

$$\text{But } a + b = 4 + 5 = 9, \text{ and } a \times b = 4 \times 5 = 20.$$

$$96 - 5 = 95 - 4 = 91$$

The answer is 9,120.

Have students practice multiplying numbers that are close to but less than 100.

Practice

Use the pattern for numbers that are close to but less than 100 to solve these problems.

1. 89
    × 99

2. 86
    × 93

3. 90
    × 79

4. 99
    × 89

Bonus problems: Have students practice finding patterns when multiplying numbers that are greater than but close to 100.

Bonus Practice

Look for a pattern in these problems. Study the answers if you can't figure out the pattern. This is challenging!

1. 103
    × 108

2. 116
    × 104

3. 125
    × 114

4. 103
    × 96

Answers to Bonus Problems

1.
$$103 \times 108$$

$100 + 3 = 103$ $3 \times 8 = 24$
$100 + 8 = 108$

103 ⤫ 3
108 ⤫ 8 $103 + 8 = 108 + 3 = 111$

Answer: 11,124

2.
$$116 \times 104$$

$100 + 16 = 116$ $4 \times 16 = 64$
$100 + 4 = 104$

116 ⤫ 16
104 ⤫ 4 $116 + 4 = 104 + 16 = 120$

Answer: 12,064

3.
$$125 \times 114$$

$100 + 25 = 125$ $14 \times 25 = 350$
$100 + 14 = 114$

125 ⤫ 25
114 ⤫ 14 $125 + 14 = 114 + 25 = 139$

Answer: 13,900
         +   350
        ─────────
         14,250 (This one was a little harder!)

4.
$$103 \times 96$$

$100 + 3 = 103$ $3 \times {}^-4 = {}^-12$
$100 - 4 = 96$

103 ⤫ +3
96 ⤫ ⁻4 $103 - 4 = 96 + 3 = 99$

Answer: 9,900
        −   12
        ─────────
         9,888

Tell the students that if they got these right, they are on their way to being great mathematicians!

Chapter Wrap-Up

Further Reteaching

This chapter uses manipulatives (two-color counters and algebra tiles) and mental math to model various algebraic operations. As demonstrated, there are algorithms for solving each of these types of problems. If students do not discover these algorithms in working through the material presented in this section, go back over the problems and encourage students to translate the patterns they see into algorithmic solutions.

Skill Maintenance and Assessment

Practice
Match each statement on the left with the equivalent statement on the right.

1. $x^2 \cdot x^2$ **a.** $x = 0$

2. $5^x = 25$ **b.** 68

3. $17^x = 1$ **c.** $-x^3 - 2x^2$

4. $3x^2 - 2x$ when $x = -3$ **d.** $-x^2$

5. $-4x^2 + 3x^2$ **e.** $x = 2$

6. $2x^2 + 3x^3 - 4x^2$ **f.** $-2x^2 + 3x^3$

7. $x^3 - 2x^2(x + 1)$ **g.** 33

8. $5x^3 + 7x^2$ when $x = 2$ **h.** x^4

Chapter 6

Word Problems

Chapter Introduction

What Are Word Problems?

Word problems are all around us in the real world. Students need to learn how to translate the words in word problems into mathematical terminology so they can solve these problems through modeling, drawing, algebraic sentences, logical thinking, and other processes.

Phases in Solving Word Problems

According to the world's greatest problem solver, Hungarian mathematician George Polya, there are five steps to solving any word problem:

1. Understand the problem.

2. Devise a plan for solving the problem.

3. Carry out the plan.

4. Look back to recheck results and reasoning.

5. Ask yourself: Is this problem similar to any other problem I have ever solved, or can I think of an easier, similar problem that I have solved that might show me a pattern to solve this problem?

There is another way to state these five steps:

1. Determine the precise question. (You cannot answer a question you don't understand.)

2. Develop a plan to find all the data you need to solve the problem.

3. Obtain the data by investigation, exploration, or experimentation (these are all basically the same thing).

4. Look for a pattern in the data.

5. Draw a conclusion (or conclusions) and express it in a concise manner, usually in the form of a formula, simple statement, or often an algorithm.

6. Verify your conclusion by producing some kind of proof that your answer is correct (we sometimes call this "checking your work").

Finally, to paraphrase one of Polya's best suggestions: If the problem is too massive to answer directly, try to reduce it to a problem that is similar but less massive.

Problem-Solving Strategies

Many word problems require specific strategies to solve them. The following strategies are specifically taught to help students solve word problems.

1. Make a table.

2. Make an organized list.

3. Look for a pattern.

4. Guess and check.

5. Draw a picture or graph.

6. Work backwards.

7. Solve a simpler problem.

Expectations for Middle School

Word problems are part of any middle school math curriculum and form a substantial portion of state and national mathematics achievement tests. Applying grade-level math skills in word problems and practicing word problems are basic requirements of any math program.

Common Problems Students Have with Word Problems

Although students have encountered word problems in math since the early grades, many students have tremendous difficulty with them throughout their schooling. The basic elements of learning word problems and the difficulties associated with each are listed following.

Word Comprehension Reads a word problem with comprehension.	*Unless students understand the words in the problem, they will be unable to solve it.*
Vocabulary Comprehension Understands math terms such as *sum*, *product*, and *equals*.	*If students do not know math vocabulary, they will not fully comprehend the question asked in the word problem.*
Identifies Questions Focuses on the question to be answered in solving the problem.	*If students look for key words and numbers but do not understand the question being asked, they cannot solve word problems meaningfully.*
Identifies Operations Identifies the operation needed to solve the problem.	*If students look for key words and numbers but do not read to identify the operation, they will solve problems incorrectly.*
Finds Relevant Information Distinguishes and separates relevant and irrelevant information based on the problem question.	*Unless students differentiate between relevant and irrelevant information, they will make significant errors when solving word problems.*
Builds Equations Translates the information in a word problem into a mathematical equation that answers the problem's question.	*If students cannot use the information from the problem to form an equation with the correct factors and relationships that focus on the problem question, they cannot solve word problems meaningfully.*
Mathematics Proficiency Understands how to use mathematics effectively and efficiently to solve word problems.	*Even if students understand the vocabulary and problem question, they cannot solve word problems unless they know the mathematics to use.*

What Research Says

Research has identified the following issues relating to children's ability to solve word problems.[1]

1. **Understanding the Question.** Many students expect word problems to follow a pattern usually presented in immediately preceding exercises. As a consequence, they often do not read word problems but simply look for numbers and treat them as the previous exercises.

2. **Extracting Relevant Data.** As mathematics becomes more complex and that complexity is reflected in word problems, many students have difficulty distinguishing between relevant and irrelevant detail. Often this is a consequence of not identifying the question being asked in the word problem.

3. **Concrete versus Abstract Mathematics.** Word problems in the higher grades appropriately involve abstract mathematics. For students who have relied on manipulatives and concrete representations to solve word problems at the lower levels, the abstraction eliminates a foundational problem-solving strategy.

4. **Factual versus Hypothetical.** Word problems in the higher grades involve hypothetical as opposed to factual questions. For students who have relied on concrete factual examples that are easily understood in real-world terms, the hypothetical nature of many word problems is distracting and difficult to comprehend.

[1] See National Research Council, *Adding It Up: Helping Children Learn Mathematics* (2001), for a more complete summary.

Model:
Find Relevant Information

Distinguishes and separates relevant and irrelevant information based on the problem question.

Model reading for relevant detail.

Solving word problems involves several skills. Two of the most important skills are being able to read the words and paying attention to details. Here are a few questions that focus on "attention to details." Let your students try these for practice.

1. You are in a race. You overtake the last person. What position are you in now?

2. You are in a race. You overtake the last person. What position *were* you in?

3. You are in a race. You overtake the second person. What position are you in?

4. Mary's father has five daughters: Nana, Nene, Nini, and Nona. What is the name of the fifth daughter?

5. A mute person wants to buy a toothbrush. By imitating the action of brushing his teeth, he successfully expresses himself to the clerk and is able to purchase the toothbrush. Now, a blind person comes into the store and wants to buy a pair of sunglasses. How should he express himself?

6. Do the following mentally.

   Take 1,000 and add 400 to it.

   Now add 30.

   Add another 1,000.

   Now add 20.

   Now add another 1,000.

   Then add 10.

   What's the total?

7. You are driving a bus. On the first stop 10 people get on. At the second stop 20 people get on and 7 get off. At the next stop, 5 people get on and 16 get off. What is the name of the bus driver?

8. If a puppy in a box weighs 2 pounds 10 ounces, and the box weighs 10 ounces, what does the puppy weigh?

9. A wire rope ladder with rungs one foot apart is hanging over the side of a boat. The bottom rung is resting on the surface of the ocean. If the tide rises at the rate of 12 inches per hour, how long will it take the first four rungs to be covered with water?

10. What is the largest number that can be made with two digits?

11. Is there a 4th of July in England?

12. How many outs are there in an inning?

13. How many birthdays does the average person have?

14. Some months have 31 days. How many have 28?

15. Is it legal for a man in California to marry his widow's sister?

16. If there are three apples and you take away two, how many do you have left?

17. A farmer has 17 sheep, and all but 9 die. How many are left?

18. How many two-cent stamps are in a dozen?

19. A doctor gives you three pills and tells you to take one every half-hour. How many minutes would the pills last?

20. Divide 30 by $\frac{1}{2}$ and add 10. What is the answer?

Model:	
The "Do Something" Method	Translates the information in a word problem into a mathematical equation that answers the problem's question.

Use problem-solving strategies to build equations: the "Do Something" method.

In many years of teaching mathematics to students and teachers, this author has seen all sorts of strategies for dealing with word problems. Unfortunately, none of them seems to work very well. Often students are instructed to "read the problem completely" and then "reread the problem." When using this approach, however, many students are still unable to say what the problem is about.

An alternative approach is to *do something* while reading a word problem. This method may call for the student to draw a picture and label it, create a chart of the data presented in the problem, write down the data, write an algebraic equation, or perform some other understandable procedure. This approach will almost always help the student know what to do, give a hint as to how to do it, and make the problem question obvious.

Have students practice "doing something."

Practice

Read each of these problems. As you read, stop and take notes, draw a picture, make a chart, or do something else as a reminder of the details of the problem.

(These problems were actually written by students!)

1. Bob bought a bike for $125.00 and then sold it for $110.00. A little later he bought it back for $90.00 and had to replace a tire for $20.00. Then he sold it for $105.00. What was his profit or loss? [Andy, age 11]

2. Maggie plays bingo three days in a row. Each day she tripled her winnings from the previous day. On the third day she won $918.00. How much did she win on her first day? [Max, age 10]

Use variables to build equations.

Algebra is THE LANGUAGE OF MATHEMATICS and is used with almost every problem one can imagine. Learn the language so you can "speak" it with fluency.

When we use variables to represent unknown information as we read through a problem, it helps us to focus on the question, find the relevant information, and identify the mathematical relationships needed to build an equation to solve the problem. There are several advantages to this approach. First, it blends algebra and arithmetic into problem solving. Second, we jot down the information as we read it; there's no need to go back over it. Third, we write algebraic equations that tell us what to do to solve the problem.

To emphasize these last two points, the motto of this author's approach to problem solving is "Write as you read." It is vitally important for students to write down the information *as they read the problem, not afterwards*—in other words, read, write, read, write, and so on.

Have students practice using variables.

Practice

Write the following symbolically (using a variable). The first three are done for you.

1. A number plus two $x + 2$

2. Three times a number less seven $3a - 7$

3. One-half the sum of a number plus eight $\frac{1}{2}(x + 8)$ or $\frac{x + 8}{2}$

4. Four more than one-half of a number

5. A number times itself plus six

6. A number added to seven equals three times that number minus five

7. One-half the sum of a number plus three equals four times that number added to two

8. The sum of two numbers

9. The difference between three times the sum of a number plus two and half of the number

10. The sum of two consecutive whole numbers

Model restating the problem with variables.

Let's look at how we can use variables to "set up" problems for solution. For now, let's focus on the solution sentences. We'll deal with the actual solutions later.

Problem 1:

Pedro has two numbers in mind. If you add them, you get 14. If you multiply them, you get 48. What are the two numbers?

Read the first sentence: *Pedro has two numbers in mind.*

Do something: Think: Pedro has two number in mind, x and y.

Write: x, y.

Read the second sentence: *If you add them, you get 14.*

Do something: Write: $x + y = 14$.

Read the third sentence: *If you multiply them, you get 48.*

Do something: Write: $(x)(y) = 48$.

Now read the question: *What are the two numbers?*

Do something: Summarize: $x + y = 14$

$(x)(y) = 48$

That's the solution "set up." We'll deal with the solution later, but for now we have a great start.

Problem 2:

Charlotte is thinking of two numbers whose sum is 16, and the difference between the two numbers is 4. What are the numbers?

Read the first part of the first sentence: *Charlotte is thinking of two numbers.*

Do something: Write a and b to represent the two numbers.

Read the second part of the first sentence: *The sum of the two numbers is 16.*

Do something: Write: $a + b = 16$

Read the third part of the first sentence: *The difference between the two numbers is 4.*

Do something: Write: $a - b = 4$

Read the question: *What are the numbers?*

Do something: Summarize: $a + b = 16$

$a - b = 4$

We will find the solution later.

Problem 3:

The Peterson family has four children who were each born two years apart. The sum of their ages is 84. What are their individual ages?

Read the first part of the first sentence: *The Peterson family has four children …*

Do something: Write *a, b, c,* and *d* to represent the four children.

Read the second part of the first sentence: *… who were born two years apart.*

Do something: Write: *a*

$$b = a + 2$$

$$c = a + 4$$

$$d = a + 6$$

Read the second sentence: *The sum of their ages is 84.*

Do something: Write: *a*

$$a + 2$$

$$a + 4$$

$$+ \quad a + 6$$

$$\overline{\qquad 84 \qquad}$$

Read the question: *What are their individual ages?*

We will find the solution later.

Problem 4:

Carl ran a 12-mile race at an average speed of 8 miles/hour. Jon ran the same race at a pace of 6 miles/hour. How many minutes longer did Jon take to complete the race than Carl? (Remember that distance = rate × time.)

Do something: As you read, fill in the data in the chart.

	Distance	Rate	Time
Carl	12	8	t_c
Jon	12	6	t_j

Read the question: *How much longer did Jon take than Carl?*

We will find the solution later.

Problem 5:

Jack is two times as old as his sister, Maggie. Together their ages add to 21. How old are they?

Read the first sentence: *Jack is two times as old as his sister Maggie.*

Do something: Write: Jack = J; Maggie = M; and $J = 2M$.

Read the second sentence: *Together their ages add to 21.*

Do something: Write: $J + M = 21$

Read the question: *How old are they?*

Do something: Summarize: $J + M = 21$; $J = 2M$.

We will find the solution later.

Problem 6:

There are 14 animals in a field. Some are chickens and some are pigs. Altogether, they have 36 legs. How many are chickens and how many are pigs?

Read the first sentence: *There are 14 animals in a field.*

Do something: Write: 14 animals

Read the second sentence: *Some are chickens and some are pigs.*

Do something: Write: Chickens = c; Pigs = p; $c + p = 14$

Read the third sentence: *Altogether, they have 36 legs.*

Do something: Think: Chickens have 2 legs, pigs have 4 legs, so …

Write: $2c + 4p = 36$

Read the question: *How many are chickens and how many are pigs?*

Do something: Summarize: $c + p = 14$; $2c + 4p = 36$

We will find the solution later.

Problem 7:

A rectangle has one side that is 4 inches longer than another. The perimeter is 32 inches. What are the dimensions of the rectangle?

Read the first sentence: *A rectangle has one side that is 4 inches longer than another.*

Do something: Draw a picture.

Read the second sentence: *The perimeter is 32 inches.*

Do something: Write an equation: $w + w + 4 + w + w + 4 = 32$

Read the question: *What are the dimensions of the rectangle?*

We will find the solution later.

Problem 8:

4 miles

3 miles

The dimensions of a rectangular field are 4 miles by 3 miles. What is the area of the field? (Remember that the area of a rectangle is length × width.)

$A = l \times w = 4$ miles $\times 3$ miles $= 12$ square miles

Problem 9:

How much interest must be paid on a loan of $4,000 at a rate of $5\frac{1}{2}$% for 2 years?

Interest = principle × rate × time or I = percent

Therefore, $I = 4,000 \times 0.055 \times 2 = \$440$

Now let's look at some possible solutions for story problems 1–9.

Problem 1: $x + y = 14$

$(x)(y) = 48$

Here's one method to try: Write the factors of 48.

x	y
48	1
24	2
12	4
8	6

Then, look for factors whose sum is 14. That's it—8 × 6.

Answer: The two numbers Pedro has in mind are 8 and 6.

(There are other ways to solve this kind of problem algebraically, but those methods are probably beyond the abilities of middle school students. However, you may want to spend some time teaching one of the methods, such as substituting, adding and subtracting (eliminating), or maybe even graphing in some cases.)

Problem 2: $a + b = 16$

$a - b = 4$

Here's one method: Let's write several sums that equal 16.

a	b
12	4
10	6
14	2
9	5

There's one that fits the problem: 10 and 6. The sum of the two numbers is 16, and the difference between them is 4.

Answer: The two numbers Charlotte is thinking of are 10 and 6.

Problem 3:

$$a$$
$$a + 2$$
$$a + 4$$
$$\underline{a + 6}$$
$$4a + 12 = 84$$
$$4a = 72$$
$$a = 18, \quad \text{so } a + 2 = 20$$
$$a + 4 = 22$$
$$a + 6 = 24$$

Check your answer: $18 + 20 + 22 + 24 = 84$

Answer: The ages of the four children are 18, 20, 22, and 24.

Problem 4:   Carl $12 = 8t_c$

$t_c = 1\frac{1}{2}$ (time in hours)

Jon $12 = 6t_j$

$t_j = 2$ (time in hours)

Answer: Jon took half an hour, or 30 minutes, longer. The problem asked for the answer to be stated in minutes, so the answer is 30 minutes.

Problem 5: $J + M = 21$

$J = 2M$

Here's one way to solve the problem: Write numbers that sum to 21 in a table, and look for $J = 2$ times M.

J	M
14	7

We got lucky the first time. That's it!

Answer: Maggie is 7 and Jack is 14.

Here's another way to solve the problem:

$J = 2M$

$J + M = 21$

Substitute $2M$ for J: $2M + M = 21$

$3M = 21$

$M = 7$

Problem 6: $c + p = 14$

                $2c + 4p = 36$

Make a chart:

Chickens		Pigs		
Number	Legs	Number	Legs	Total Legs
6	12	8	32	44 no
8	16	6	24	40 no
9	18	5	20	38 no
12	24	4	16	40 no
10	20	4	16	36 yes

Answer: There are 10 chickens and 4 pigs in the field.

Problem 7: $w + w + 4 + w + w + 4 = 32$

$4w + 8 = 32$

$4w = 24$

$w = 6$

$w + 4 = 10$

Answer: The dimensions of the rectangle are 6 inches by 10 inches.

Problem 8: Already solved by formula.

Answer: 12 square miles

Problem 9: Already solved by formula.

Answer: $440.00

Have students practice writing variables to represent information in word problems.

<table>
<tr><td>**Practice**</td></tr>
</table>

Write the following sentences in the language of algebra using variables.

1. The area of a rectangle equals length times width.

2. The perimeter of a rectangle equals two times the length plus two times the width.

3. The area of a triangle equals one-half the base times the height.

4. To change feet to inches, multiply the number of feet by 12.

5. The volume of a cube equals the length of a side times three or the length cubed.

6. To convert feet to yards, multiply the number of feet by $\frac{1}{3}$ (or divide the number of feet by 3).

Chapter Wrap-Up

Assess students.

Here are some more problems written by children ages 9 to 14. Have students use their new skills with word problems to solve them.

1. The Wright, Winber, and Wallace families exchange presents each year. Each family buys a present for every member of the other two families. The Wallace family bought 10 presents, the Wright family bought 11 presents, and the Winber family bought 13 presents. How many people are in each family? [by Max, age 10]

2. A bus driver started out with 20 people on the bus. At the first stop, 10 people got off and 4 got on. At the second stop, 11 got off and 7 got on. At the next stop, 9 got off and 11 got on. Bus fare is 50¢. How much fare money did the bus driver collect? How many people were on the bus at the end? [by Chris, age 9]

3. There were 3 people on the bus. Each time the bus stopped, two more than the first number got on. How many people were on the bus after the sixth stop? [by Kate, age 9]

Further Reteaching

This chapter examines various strategies for helping students solve word problems, including looking for details and doing something as you read—such as drawing a picture, making a chart, or writing variables.

The best teaching practice for helping students with word problems is to pose interesting, relevant problems and give students time and support to solve them. But do not *show* students how to solve these problems.

The most important element in solving word problems is reasoning and employing all you know about words, math words, and mathematics to figure out the puzzle. If a student is shown how to solve a word problem instead of allowed to reason through it, with supportive questions to help, it will rob the student of the best possible problem-solving practice, which is solving word problems!

Skill Maintenance and Assessment

A. Have students solve these problems using the "do something" method—jotting down information *as they read*.

1. As camp counselors, Madeline earns $\frac{1}{3}$ more per week than Becca because she works with the younger children. If Becca's weekly salary is $240, what is Madeline's weekly salary?

2. The number of boys in a homeroom class is seven times the number of girls. The number of boys is 28. How many girls are in the class?

Solutions follow:

1. Let B = Becca, M = Madeline

   $M = B + \frac{1}{3} B$

   $M = 240 + 80 = 320$

   Answer: Madeline makes $320 and Becca makes $240—one third less than Madeline.

2. Let B = boys, G = girls

   $B = 7G$

   $28 = 7G$

   $G = 4$

   Answer: There are 4 girls in the class.

B. Have students write their own word problems for other students to solve. **(Research has shown that students who construct their own problems become better problem solvers.)**

Answer Key

Chapter 1

Chapter Wrap-Up:
Practice Ideas

Algorithm Practice
Page 15

   1. 442

   2. 6,312

Chapter 2

Model:
Division in Arrays

Assess division.
Page 27

   1. 10

   2. 8

   3. 9

   4. 14

   5. 9

Model:
Standard Algorithm in Long Division

Practice rounding.
Page 29

   1. 1,200

   2. 3,460

   3. 8,000

   4. 33

   5. 40

   6. 12.4

Practice using compatible numbers.
Page 30
Answers will vary. They should be reasonably close to the actual values:

   1. 3,703

   2. 6,776

   3. 1,961

   4. 12,040

   5. 83,334

   6. 22,400

7. 6,825

8. 30,492

Practice multiplying by multiples of 10.
Page 31

1. 1,000 or 10^3

2. 10,000 or 10^4

3. 10,000,000 or 10^7

4. 10,000,000 or 10^7

5. 10,000,000 or 10^7

Practice estimating.
Page 32
Answers will vary. They should be reasonably close to the actual values:

1. 501

2. 5,160

3. 2,052

4. 866

5. 4.111

Practice long division.
Page 34

1. 41 r.6

2. 120 r.4

3. 138 r.1

4. 165 r.2

5. 66 r.1

6. 86 r.4

7. 145 r.4

8. 116 r.3

9. 106 r.1

10. 52

11. 110

12. 114 r.10

13. 182 r.2

Practice division with two-digit divisors less than 20.
Page 36

1. 152 r.5

2. 182 r.1

3. 138 r.2

4. 717

Practice long division.
Page 39

1. 30 r.47

2. 32 r.13

3. 115

4. 45 r.7

5. 68

6. 139 r.54

Chapter Wrap-Up:
Extra Practice Ideas
Page 39
Answers will vary. They should be reasonably close to the actual values:

1. 22.1

2. 216.88

3. 309.27

4. 303.06

5. 1,387.02

Inverse Operations Practice
Page 40

1. $12 \div 4 = 3$ or $12 \div 3 = 4$
2. $84 \div 7 = 12$ or $84 \div 12 = 7$
3. $600 \div 25 = 24$ or $600 \div 24 = 25$
4. $256 \div 16 = 16$

Algorithm Practice
Page 40

1. 10.96
2. 26.6
3. 3.4
4. 2.6
5. 3

Chapter 3

Practice modeling different fractions of the whole.
Page 48

1. whole 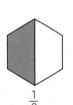 $\frac{1}{2}$

2. whole  $\frac{1}{3}$

3. 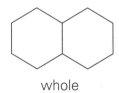 whole 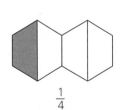 $\frac{1}{4}$

4. whole $\frac{2}{3}$

5. whole  $\frac{4}{6}$

6. 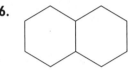 whole 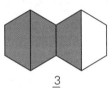 $\frac{3}{4}$

7. whole $\frac{3}{2}$

8. whole $\frac{4}{3}$

9. whole 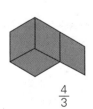 $\frac{1}{6}$

10. whole 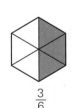 $\frac{3}{6}$

11.

whole

$\frac{5}{6}$

12.

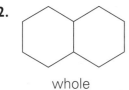

whole

$\frac{5}{12}$

Practice using < (less than) and > (greater than) signs.

Page 49

1. <
2. <
3. <
4. <
5. <
6. >
7. <
8. >
9. >
10. <
11. >

Model:

Multiplying Fractions by Whole Numbers

Practice multiplying fractions by whole numbers.

Page 52

1. C
2. D
3. E
4. B
5. F
6. A

Model:

Simplifying Fractions

Assess simplifying fractions.

Page 63

1. $\frac{5}{6}$

Model:

Adding and Subtracting Fractions with Common Denominators

Assess subtracting fractions.

Page 65

1. $\frac{2}{5}$

Model:
Adding Fractions with Unlike Denominators

Assess adding fractions with unlike denominators.
Page 68

Denominators must be the same. If denominators are not the same, find the lowest common denominator. Then, add the numerators but not the denominators.

Chapter Wrap-Up:
Adding Fractions
Page 73

1a. $\frac{19}{84}$ b. $\frac{22}{15}$ c. $\frac{5}{12}$

2a. $\frac{13}{22}$ b. $\frac{3}{2}$ c. $\frac{115}{77}$

3a. 1 b. $\frac{29}{24}$ c. $\frac{5}{6}$

4a. $\frac{85}{77}$ b. $\frac{57}{56}$ c. $\frac{5}{6}$

Subtracting Fractions
Page 73

1a. $\frac{45}{88}$ b. $\frac{1}{18}$ c. $\frac{2}{3}$

2a. $\frac{7}{66}$ b. $\frac{2}{5}$ c. $\frac{1}{42}$

3a. $\frac{1}{42}$ b. $\frac{1}{12}$ c. $\frac{3}{40}$

4a. $\frac{4}{11}$ b. $\frac{1}{20}$ c. $\frac{1}{15}$

Multiplying Fractions
Page 74

1a. $\frac{3}{14}$ b. $\frac{1}{3}$ c. $\frac{5}{42}$

2a. $\frac{4}{9}$ b. $\frac{7}{18}$ c. $\frac{1}{10}$

3a. $\frac{8}{66}$ b. $\frac{3}{32}$ c. $\frac{1}{28}$

4a. $\frac{1}{6}$ b. $\frac{1}{8}$ c. $\frac{80}{99}$

Dividing Fractions
Page 74

1a. $\frac{5}{12}$ b. $\frac{28}{11}$ c. $\frac{1}{2}$

2a. $\frac{11}{54}$ b. 3 c. $\frac{36}{7}$

3a. $\frac{11}{20}$ b. $\frac{1}{3}$ c. $\frac{42}{55}$

4a. $\frac{5}{6}$ b. $\frac{3}{4}$ c. $\frac{1}{2}$

Chapter 4

Model:
Place Value

Practice decimal fraction place value.
Page 80

1,000s	100s	10s	1s	.	Tenths	Hundredths	Thousandths
		1	3	.	2		
		1	3	.	3	4	
5	2	8	9	.	2		
9	8	4	3	.	2	3	1
				.	0	0	1

Assess reading decimal fractions.
Page 81

1. 124.532

2. 200.002

3. 1,454.5

4. 222.22

5. 1,000.001

Practice finding equivalent decimals.
Page 83

1. 0.5   **6.** 0.12

2. 3.6   **7.** 0.03

3. 5.4   **8.** 0.0066

4. 10.05   **9.** 0.01

5. 0.8   **10.** 4.3

Model:
Number Line

Practice plotting decimals and decimal fractions on a number line.
Page 84

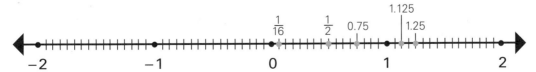

Model:
Operations with Decimals

Practice adding and subtracting decimals.
Page 85

1. 5.66
2. 7.17
3. 10.51
4. 13.106
5. 2.51
6. 3.11
7. 4.81
8. 4.52

Practice multiplying decimals.
Page 86

1. 0.72
2. 1.5531
3. 16.2
4. 85.4
5. 0.0675
6. 0.4024

Practice dividing decimals.
Page 88

1. 20.83
2. 182.47
3. 2.229
4. 219.5
5. 0.00637
6. 30.19

Assess computing decimals.
Page 88

1. 1.22
2. 3.38
3. 5.82
4. 0.78
5. 0.336
6. 2.484
7. 11.034
8. 3.6

Chapter Wrap-Up:
Practice
Page 90

1. −0.648
2. 0.772
3. −2.667
4. 2.973
5. 0.11
6. 9
7. 1.4
8. 5,477.675
9. 0.4
10. 0.5
11. 1.1
12. 0.24
13. −0.5
14. 104.1
15. 9.372
16. 16.208
17. 29.05

18. 4.4

19. 2,393.82

20. 239.382

21. 5.7

22. 266

23. 6.342

24. 6.458

25. 3.337

26. 5.78

27. 8.84

Chapter 5

Model:
Addition, Subtraction, Multiplication, and Division with Two-Color Counters

Practice adding positive and negative integers.
Page 100

1. 8 **2.** −2 **3.** 2

4. −8 **5.** 6 **6.** −6

7. 2

Practice subtracting positive and negative integers.
Page 101

1. 9 **2.** −1 **3.** −5

4. −6 **5.** −5 **6.** 6

7. −1 **8.** −2 **9.** 5

10. 6

Practice multiplying and dividing integers.
Page 103

1. 16 **2.** −16 **3.** 16

4. −1 **5.** 1 **6.** −1

7. 1

Model:
Exponents (moving from patterns to rules)

Practice with exponents.
Page 105

1. $\frac{1}{4}$ or 4^{-1} **2.** 4 or 4^{1}

3. 1 or 4^{0} **4.** $\frac{1}{a^{1}}$

5. 1 or a^{0}

Practice describing patterns in words and symbols.
Page 106

1. **a.** 29, 47, 76
   b. Add each number to the previous number to get the next number in the series.
   c. $a + b = c,\ b + c = d,\ d + c = e$

2. **a.** 64, 128, 256
   b. Double previous number.
   c. $2a = b,\ 2b = c,\ 2c = d$

3. **a.** −6; add 7; $x + 7 = y$
   b. −7; subtract 7; $x − 7 = y$
   c. 10; multiply by 3 and add 1; $3x + 1 = y$

Practice translating words into symbols and symbols into words.
Pages 106–107

1. **a.** $3x^2$
   b. $8 + 3x$
   c. $11(x + 6)$
   d. $4 + (x - 6)$
   e. $x + (x + 4) = 24$

2. **a.** A number plus nine equals sixteen
   b. Three times the sum of a number plus two less six equals twenty-four
   c. Four times a number plus two equals seven
   d. Three times a number less seven equals thirteen

Practice simplifying expressions.
Page 108

1. $6x^2$ 2. $2 + \frac{1}{2}x$

3. $-2x - 7$ 4. $16x$

5. $5x + 4$ 6. $\frac{1}{4}$

7. $\frac{1}{12}y + 4$ 8. $2x$

9. $6x + 15y$ 10. $5x + y$

Practice solving linear equations.
Page 112

1. $x = -8$ 2. $x = -5$

3. $x = \frac{8}{3}$ 4. $x = -1$

Practice solving linear equations.
Page 113

1. 3   2. $-\frac{8}{3}$ 3. 0

4. 0   5. $\frac{4}{3}$ 6. 1

7. $\frac{1}{5}$ 8. $\frac{13}{5}$ 9. $\frac{28}{9}$

10. 3   11. -3 12. $\frac{10}{8}$

13. $\frac{29}{7}$ 14. $-\frac{17}{4}$ 15. $-\frac{1}{2}$

16. 7   17. $-\frac{20}{11}$ 18. 1

19. $\frac{3}{2}$ 20. $\frac{1}{8}$

Practice modeling polynomials.
Page 116.

1.

2.

3.

4.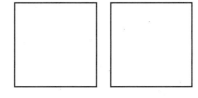

9.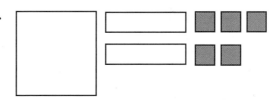

10.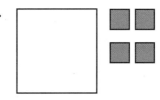

5.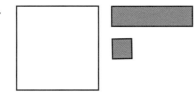

6.

7.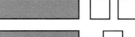

8.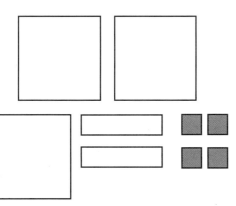

Practice adding polynomials without using algebra tiles.
Page 118

1. $4x^2 - x + 3$ **2.** $3x^2 - 2x - 3$

3. $2x^2 + 6x + 1$ **4.** $2x^2 + x + 1$

5. $2x^2 + x - 5$ **6.** $2x^2 - x + 1$

7. $2x^2 - 3x + 1$ **8.** $3x^2 + 2x - 5$

Practice subtracting polynomials.
Page 122

1. $2x^2 - 3x + 4$ **2.** $3x^2 + 3x - 4$

3. $3x^2 - 2x + 3$ **4.** $x^2 + 7x + 5$

5. $-2x$ **6.** $x^2 - x - 3$

7. $x^2 - 2x + 7$

Practice multiplying polynomials.
Page 124

1. $x^2 + x$ **2.** $x^2 + 4x + 3$

3. $2x^2 + 7x + 3$ **4.** $x^2 + 4x + 4$

5. $x^2 + 5x + 4$ **6.** $2x^2 + 8x + 6$

Practice multiplying polynomials with negative numbers.
Page 127

1. $x^2 - 9$

2. $x^2 - 4x + 4$

3. $3x^2 + 2x - 8$

4. $x^2 - 4x + 3$

5. $x^2 - 7x + 12$

6. $2x^2 + 5x - 3$

Assess dividing and factoring trinomials.
Page 131

1. $(x + 3)(x + 2)$ 2. $(x - 2)(x + 1)$

3. $(2x - 1)(x + 3)$ 4. $(x - 2)(x + 2)$

5. $(2x - 3)(2x + 3)$ 6. $(x - 4)(x + 4)$

7. $(x + 3)$ 8. $(x + 1)$

9. $(x + 4)$ 10. $(x - 3)$

Model:
Mental Math

Practice multiplying by eleven using mental math.
Page 133

1. 374

2. 297

3. 253

4. 418

5. 4,675

Practice squaring numbers ending in 5.
Page 135

1. 4,225

2. 5,625

3. 7,225

4. 9,025

Practice squaring numbers not ending in 5.
Page 135

1. 1,156

2. 729

3. 2,809

4. 5,476

5. 7,569

Practice mental math.
Page 137

1. 143

2. 399

3. 899

4. 249,999

5. 999,999

Practice mental math.
Page 138

1. 616

2. 1,224

3. 2,009

4. 3,021

5. 4,216

Practice multiplying numbers that are close to but less than 100.
Page 141

   1. 8,811

   2. 7,998

   3. 7,110

   4. 8,811

Practice skill maintenance and assessment.
Page 143

   1. h   **2.** e

   3. a   **4.** g

   5. d   **6.** f

   7. c   **8.** b

Chapter 6

Model:
Find Relevant Information

Practice reading for relevant detail.
Pages 150–151

   1. Second-to-last

   2. Last place

   3. Second

   4. Mary

   5. Ask for it.

   6. 3,460

   7. (Your name)

   8. 2 pounds 10 ounces

   9. Never; boats rise with the tides

  10. 9^9

  11. Yes

  12. 6

  13. 1

  14. All of them

  15. No, he's dead

  16. 2

  17. 9

  18. 12

  19. 60 minutes

  20. 70

Practice "doing something."
Page 152

   1. Bob lost $20.00.

   2. W = 102

Practice using variables.
Page 153

4. $4 + (\frac{1}{2}x)$

5. $x^2 + 6$

6. $x + 7 = 3x - 5$

7. $\frac{1}{2}(x + 3) = 4x + 2$

8. $x + y$

9. $3(x + 2) - \frac{1}{2}x$

10. $x + (x + 1)$

Practice writing variables to represent information in word problems.
Page 163

1. $a = l \times w$

2. $P = 2l + 2w$

3. $A = \frac{1}{2}bh$

4. $i = 12f$

5. $V = l^3$

6. $\frac{1}{3}f = y$

Assess solving word problems.
Page 164

1. Wrights = 6
   Winbers = 4
   Wallaces = 7

2. $21; 12 people left

3. 33